I0066430

Fertilizer Technology and Soil Fertility

Fertilizer Technology and Soil Fertility

Edited by
Virginia Munn

Larsen & Keller
www.larsen-keller.com

Fertilizer Technology and Soil Fertility
Edited by Virginia Munn
ISBN: 978-1-63549-643-7 (Hardback)

© 2018 Larsen & Keller

Larsen & Keller

Published by Larsen and Keller Education,
5 Penn Plaza,
19th Floor,
New York, NY 10001, USA

Cataloging-in-Publication Data

Fertilizer technology and soil fertility / edited by Virginia Munn.
   p. cm.
Includes bibliographical references and index.
ISBN 978-1-63549-643-7
1. Fertilizers. 2. Soil fertility. I. Munn, Virginia.
S633 .F47 2018
631.8--dc23

This book contains information obtained from authentic and highly regarded sources. All chapters are published with permission under the Creative Commons Attribution Share Alike License or equivalent. A wide variety of references are listed. Permissions and sources are indicated; for detailed attributions, please refer to the permissions page. Reasonable efforts have been made to publish reliable data and information, but the authors, editors and publisher cannot assume any responsibility for the vailidity of all materials or the consequences of their use.

Trademark Notice: All trademarks used herein are the property of their respective owners. The use of any trademark in this text does not vest in the author or publisher any trademark ownership rights in such trademarks, nor does the use of such trademarks imply any affiliation with or endorsement of this book by such owners.

For more information regarding Larsen and Keller Education and its products, please visit the publisher's website www.larsen-keller.com

Table of Contents

Preface

Fertilizers are naturally occurring materials that are used on soil to provide them with essential nutrients to ensure the proper growth of plants and crops. They can also be of synthetic nature. The most commonly known fertilizers are potassium fertilizers, nitrogen fertilizers and phosphorus fertilizers. Most of the topics introduced in this textbook cover applications of fertilizers. It provides in-depth information on the proper use of fertilizers and any threats they pose to the environment. It discusses their usefulness and importance in agricultural production. This text is appropriate for those seeking detailed information in this area.

To facilitate a deeper understanding of the contents of this book a short introduction of every chapter is written below:

Chapter 1- Fertilizers are materials used to aid plant growth. The main nutrients that they provide plants with are phosphorus, potassium and nitrogen. Fertilizers provide nutrients as well as improve soil quality. This is an introductory chapter which will introduce briefly all the significant aspects of fertilizers.

Chapter 2- Nitrogen fertilizers are produced from ammonia and can be injected into the ground directly. Ammonia is a colorless gas which is a compound of hydrogen and nitrogen. The diverse applications of nitrogen fertilizers in the current scenario have been thoroughly discussed in this chapter.

Chapter 3- Phosphorous is a macronutrient and an important fertilizer. Phosphate rock is the main source of phosphate. The chapter closely examines the key concepts of phosphorus fertilizers to provide an extensive understanding of the subject.

Chapter 4- Potassium is an important crop nutrient. It improves yield, water retention, taste and disease resistance in food crops. Potassium can be applied to sugar, corn, wheat, rise, vegetables and fruits. The chapter on potash and potassium fertilizers offers an insightful focus, keeping in mind the complex subject matter.

Chapter 5- The impact of fertilizers on the environment depends on the agronomic practices of the area. Some of the environmental issues related to the issue of the environment are deforestation, irrigation problems, soil degradation and waste. Eutrophication, soil contamination, soil salinity and human impact on the nitrogen cycle are some of the topic discussed in the following chapter. Fertilizers are best understood in confluence with the major topics listed in the following chapter.

Finally, I would like to thank the entire team involved in the inception of this book for their valuable time and contribution. This book would not have been possible without their efforts. I would also like to thank my friends and family for their constant support.

Editor

Fertilizers and its Classification

Fertilizers are materials used to aid plant growth. The main nutrients that they provide plants with are phosphorus, potassium and nitrogen. Fertilizers provide nutrients as well as improve soil quality. This is an introductory chapter which will introduce briefly all the significant aspects of fertilizers.

Fertilizer

A fertilizer is a material that furnishes one or more of the chemical elements necessary for the proper development and growth of plants. The most important fertilizers are fertilizer products (also called chemical or mineral fertilizers), manures, and plant residues. A fertilizer product is a material produced by industrial processes with the specific purpose of being used as a fertilizer. Fertilizers are essential in today's agricultural system to replace the elements extracted from the soil in the form of food and other agricultural products.

A fertilizer or fertiliser is any material of natural or synthetic origin (other than liming materials) that is applied to soils or to plant tissues (usually leaves) to supply one or more plant nutrients essential to the growth of plants.

Mechanism

Fertilizers enhance the growth of plants. This goal is met in two ways, the traditional one being additives that provide nutrients. The second mode by which some fertilizers act is to enhance the effectiveness of the soil by modifying its water retention and aeration. Fertilizers typically provide, in varying proportions:

- three main macronutrients:
  - Nitrogen (N): leaf growth;
  - Phosphorus (P): Development of roots, flowers, seeds, fruit;
  - Potassium (K): Strong stem growth, movement of water in plants, promotion of flowering and fruiting;
- three secondary macronutrients: calcium (Ca), magnesium (Mg), and sulfur (S);
- micronutrients: copper (Cu), iron (Fe), manganese (Mn), molybdenum (Mo), zinc (Zn), boron (B), and of occasional significance there are silicon (Si), cobalt (Co), and vanadium (V) plus rare mineral catalysts.

The nutrients required for healthy plant life are classified according to the elements, but the elements are not used as fertilizers. Instead compounds containing these elements are the basis of fertilizers. The macronutrients are consumed in larger quantities and are present in plant tissue in quantities from 0.15% to 6.0% on a dry matter (DM) (0% moisture) basis. Plants are made up of four main elements: hydrogen, oxygen, carbon, and nitrogen. Carbon, hydrogen and oxygen are widely available as water and carbon dioxide. Although nitrogen makes up most of the atmosphere, it is in a form that is unavailable to plants. Nitrogen is the most important fertilizer since nitrogen is present in proteins, DNA and other components (e.g., chlorophyll). To be nutritious to plants, nitrogen must be made available in a "fixed" form. Only some bacteria and their host plants (notably legumes) can fix atmospheric nitrogen (N_2) by converting it to ammonia. Phosphate is required for the production of DNA and ATP, the main energy carrier in cells, as well as certain lipids.

Micronutrients are consumed in smaller quantities and are present in plant tissue on the order of parts-per-million (ppm), ranging from 0.15 to 400 ppm DM, or less than 0.04% DM. These elements are often present at the active sites of enzymes that carry out the plant's metabolism. Because these elements enable catalysts (enzymes) their impact far exceeds their weight percentage.

Classification

Fertilizers are classified in several ways. They are classified according to whether they provide a single nutrient (e.g., K, P, or N), in which case they are classified as "straight fertilizers." "Multinutrient fertilizers" (or "complex fertilizers") provide two or more nutrients, for example N and P. Fertilizers are also sometimes classified as inorganic versus organic. Inorganic fertilizers exclude carbon-containing materials except ureas. Organic fertilizers are usually (recycled) plant- or animal-derived matter. Inorganic are sometimes called synthetic fertilizers since various chemical treatments are required for their manufacture.

Single Nutrient ("straight") Fertilizers

The main nitrogen-based straight fertilizer is ammonia or its solutions. Ammonium nitrate (NH_4NO_3) is also widely used. Urea is another popular source of nitrogen, having the advantage that it is solid and non-explosive, unlike ammonia and ammonium nitrate, respectively. A few percent of the nitrogen fertilizer market (4% in 2007) has been met by calcium ammonium nitrate ($Ca(NO_3)_2 \cdot NH_4NO_3 \cdot 10H_2O$).

The main straight phosphate fertilizers are the superphosphates. "Single superphosphate" (SSP) consists of 14–18% P_2O_5, again in the form of $Ca(H_2PO_4)_2$, but also phosphogypsum ($CaSO_4 \cdot 2 H_2O$). Triple superphosphate (TSP) typically consists of 44-48% of P_2O_5 and no gypsum. A mixture of single superphosphate and triple superphosphate is called double superphosphate. More than 90% of a typical superphosphate fertilizer is water-soluble.

Multinutrient Fertilizers

These fertilizers are the most common. They consist of two or more nutrient components.

Binary (NP, NK, PK) Fertilizers

Major two-component fertilizers provide both nitrogen and phosphorus to the plants. These are called NP fertilizers. The main NP fertilizers are monoammonium phosphate (MAP) and diammonium phosphate (DAP). The active ingredient in MAP is $NH_4H_2PO_4$. The active ingredient in DAP is $(NH_4)_2HPO_4$. About 85% of MAP and DAP fertilizers are soluble in water.

NPK Fertilizers

NPK fertilizers are three-component fertilizers providing nitrogen, phosphorus, and potassium.

NPK rating is a rating system describing the amount of nitrogen, phosphorus, and potassium in a fertilizer. NPK ratings consist of three numbers separated by dashes (e.g., 10-10-10 or 16-4-8) describing the chemical content of fertilizers. The first number represents the percentage of nitrogen in the product; the second number, P_2O_5; the third, K_2O. Fertilizers do not actually contain P_2O_5 or K_2O, but the system is a conventional shorthand for the amount of the phosphorus (P) or potassium (K) in a fertilizer. A 50-pound (23 kg) bag of fertilizer labeled 16-4-8 contains 8 lb (3.6 kg) of nitrogen (16% of the 50 pounds), an amount of phosphorus equivalent to that in 2 pounds of P_2O_5 (4% of 50 pounds), and 4 pounds of K_2O (8% of 50 pounds). Most fertilizers are labeled according to this N-P-K convention, although Australian convention, following an N-P-K-S system, adds a fourth number for sulfur.

Micronutrients

The main micronutrients are molybdenum, zinc, and copper. These elements are provided as water-soluble salts. Iron presents special problems because it converts to insoluble (bio-unavailable) compounds at moderate soil pH and phosphate concentrations. For this reason, iron is often administered as a chelate complex, e.g., the EDTA derivative. The micronutrient needs depend on the plant. For example, sugar beets appear to require boron, and legumes require cobalt.

Production

Nitrogen Fertilizers

Top users of nitrogen-based fertilizer		
Country	Total N use (Mt pa)	Amt. used for feed/pasture (Mt pa)
China	18.7	3.0
India	11.9	N/A
U.S.	9.1	4.7
France	2.5	1.3
Germany	2.0	1.2
Brazil	1.7	0.7

Canada	1.6	0.9
Turkey	1.5	0.3
UK	1.3	0.9
Mexico	1.3	0.3
Spain	1.2	0.5
Argentina	0.4	0.1

Nitrogen fertilizers are made from ammonia (NH_3), which is sometimes injected into the ground directly. The ammonia is produced by the Haber-Bosch process. In this energy-intensive process, natural gas (CH_4) supplies the hydrogen, and the nitrogen (N_2) is derived from the air. This ammonia is used as a feedstock for all other nitrogen fertilizers, such as anhydrous ammonium nitrate (NH_4NO_3) and urea ($CO(NH_2)_2$).

Deposits of sodium nitrate ($NaNO_3$) (Chilean saltpeter) are also found in the Atacama desert in Chile and was one of the original (1830) nitrogen-rich fertilizers used. It is still mined for fertilizer.

There has been technical work investigating on-site (on-farm) synthesis of nitrate fertilizer using solar photovoltaic power, which would enable farmers more control in soil fertility, while using far less surface area than conventional organic farming for nitrogen fertilizer.

Phosphate Fertilizers

All phosphate fertilizers are obtained by extraction from minerals containing the anion PO_4^{3-}. In rare cases, fields are treated with the crushed mineral, but most often more soluble salts are produced by chemical treatment of phosphate minerals. The most popular phosphate-containing minerals are referred to collectively as phosphate rock. The main minerals are fluorapatite $Ca_5(PO_4)_3F$ (CFA) and hydroxyapatite $Ca_5(PO_4)_3OH$. These minerals are converted to water-soluble phosphate salts by treatment with sulfuric or phosphoric acids. The large production of sulfuric acid as an industrial chemical is primarily due to its use as cheap acid in processing phosphate rock into phosphate fertilizer. The global primary uses for both sulfur and phosphorus compounds relate to this basic process.

In the nitrophosphate process or Odda process (invented in 1927), phosphate rock with up to a 20% phosphorus (P) content is dissolved with nitric acid (HNO_3) to produce a mixture of phosphoric acid (H_3PO_4) and calcium nitrate ($Ca(NO_3)_2$). This mixture can be combined with a potassium fertilizer to produce a *compound fertilizer* with the three macronutrients N, P and K in easily dissolved form.

Potassium Fertilizers

Potash is a mixture of potassium minerals used to make potassium (chemical symbol: K) fertilizers. Potash is soluble in water, so the main effort in producing this nutrient from the ore involves some purification steps; e.g., to remove sodium chloride (NaCl) (common salt). Sometimes potash is referred to as K_2O, as a matter of convenience to those describing the potassium content. In fact potash fertilizers are usually potassium chloride, potassium sulfate, potassium carbonate, or potassium nitrate.

Compound Fertilizers

Compound fertilizers, which contain N, P, and K, can often be produced by mixing straight fertilizers. In some cases, chemical reactions occur between the two or more components. For example, monoammonium and diammonium phosphates, which provide plants with both N and P, are produced by neutralizing phosphoric acid (from phosphate rock) and ammonia :

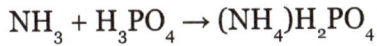

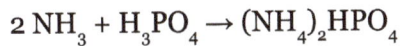

$$NH_3 + H_3PO_4 \rightarrow (NH_4)H_2PO_4$$
$$2\,NH_3 + H_3PO_4 \rightarrow (NH_4)_2HPO_4$$

Organic Fertilizers

Compost bin for small-scale production of organic fertilizer

A large commercial compost operation

The main "organic fertilizers" are peat, animal wastes, plant wastes from agriculture, and treated sewage sludge (biosolids). In terms of volume, peat is the most widely used organic fertilizer. This immature form of coal confers no nutritional value to the plants, but improves the soil by aeration and absorbing water. Animal sources include the products of the slaughter of animals. Bloodmeal, bone meal, hides, hoofs, and horns are typical components. Organic fertilizer usually contain fewer nutrients, but offer other advantages as well as being appealing to those who are trying to practice "environmentally friendly" farming.

Other Elements: Calcium, Magnesium, and Sulfur

Calcium is supplied as superphosphate or calcium ammonium nitrate solutions.

Application

Fertilizers are commonly used for growing all crops, with application rates depending on the soil fertility, usually as measured by a soil test and according to the particular crop. Legumes, for example, fix nitrogen from the atmosphere and generally do not require nitrogen fertilizer.

Liquid Vs Solid

Fertilizers are applied to crops both as solids and as liquid. About 90% of fertilizers are applied as solids. Solid fertilizer is typically granulated or powdered. Often solids are available as prills, a solid globule. Liquid fertilizers comprise anhydrous ammonia, aqueous solutions of ammonia, aqueous solutions of ammonium nitrate or urea. These concentrated products may be diluted with water to form a concentrated liquid fertilizer (e.g., UAN). Advantages of liquid fertilizer are its more rapid effect and easier coverage. The addition of fertilizer to irrigation water is called "fertigation".

Slow and Controlled release Fertilizers

Slow- and controlled-release involve only 0.15% (562,000 tons) of the fertilizer market (1995). Their utility stems from the fact that fertilizers are subject to antagonistic processes. In addition to their providing the nutrition to plants, excess fertilizers can be poisonous to the same plant. Competitive with the uptake by plants is the degradation or loss of the fertilizer. Microbes degrade many fertilizers, e.g., by immobilization or oxidation. Furthermore, fertilizers are lost by evaporation or leaching. Most slow-release fertilizers are derivatives of urea, a straight fertilizer providing nitrogen. Isobutylidenediurea ("IBDU") and urea-formaldehyde slowly convert in the soil to free urea, which is rapidly uptaken by plants. IBDU is a single compound with the formula $(CH_3)_2CHCH(NHC(O)NH_2)_2$ whereas the urea-formaldehydes consist of mixtures of the approximate formula $(HOCH_2NHC(O)NH)_nCH_2$.

Besides being more efficient in the utilization of the applied nutrients, slow-release technologies also reduce the impact on the environment and the contamination of the subsurface water. Slow-release fertilizers (various forms including fertilizer spikes, tabs, etc.) which reduce the problem of "burning" the plants due to excess nitrogen. Polymer coating of fertilizer ingredients gives tablets and spikes a 'true time-release' or 'staged nutrient release' (SNR) of fertilizer nutrients.

Controlled release fertilizers are traditional fertilizers encapsulated in a shell that degrades at a specified rate. Sulfur is a typical encapsulation material. Other coated products use thermoplastics (and sometimes ethylene-vinyl acetate and surfactants, etc.) to produce diffusion-controlled release of urea or other fertilizers. "Reactive Layer Coating" can produce thinner, hence cheaper, membrane coatings by applying reactive monomers simultaneously to the soluble particles. "Multicote" is a process applying layers of low-cost fatty acid salts with a paraffin topcoat.

Foliar Application

Foliar fertilizers are applied directly to leaves. The method is almost invariably used to apply water-soluble straight nitrogen fertilizers and used especially for high value crops such as fruits.

Fertilizer burn

Chemicals that Affect Nitrogen Uptake

Various chemicals are used to enhance the efficiency of nitrogen-based fertilizers. In this way farmers can limit the polluting effects of nitrogen run-off. Nitrification inhibitors (also known as nitrogen stabilizers) suppress the conversion of ammonia into nitrate, an anion that is more prone to leaching. 1-Carbamoyl-3-methylpyrazole (CMP), dicyandiamide, and nitrapyrin (2-chloro-6-tri-chloromethylpyridine) are popular. Urease inhibitors are used to slow the hydrolytic conversion of urea into ammonia, which is prone to evaporation as well as nitrification. The conversion of urea to ammonia catalyzed by enzymes called ureases. A popular inhibitor of ureases is N-(n-butyl) thiophosphoric triamide (NBPT).

Overfertilization

Careful fertilization technologies are important because excess nutrients can be as detrimental. Fertilizer burn can occur when too much fertilizer is applied, resulting in drying out of the leaves and damage or even death of the plant. Fertilizers vary in their tendency to burn roughly in accordance with their salt index.

Statistics

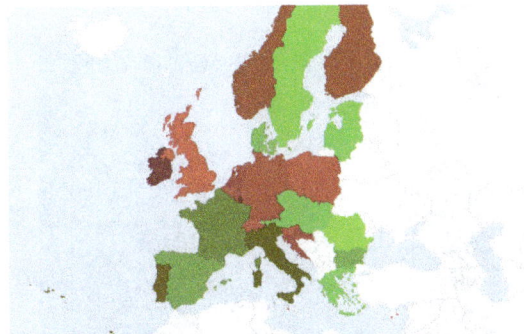

The map displays the statistics of fertilizer consumption in western and central European counties.

Conservative estimates report 30 to 50% of crop yields are attributed to natural or synthetic commercial fertilizer. Global market value is likely to rise to more than US$185 billion until 2019. The European fertilizer market will grow to earn revenues of approx. €15.3 billion in 2018.

Data on the fertilizer consumption per hectare arable land in 2012 are published by The World Bank. For the diagram below values of the European Union (EU) countries have been extracted and are presented as kilograms per hectare (pounds per acre). The total consumption of fertilizer in the EU is 15.9 million tons for 105 million hectare arable land area (or 107 million hectare arable land according to another estimate). This figure equates to 151 kg of fertilizers consumed per ha arable land on average for the EU countries. Interestingly, mainly in those countries where fertilizers are consumed a lot also plant growth product are sold more than in others.

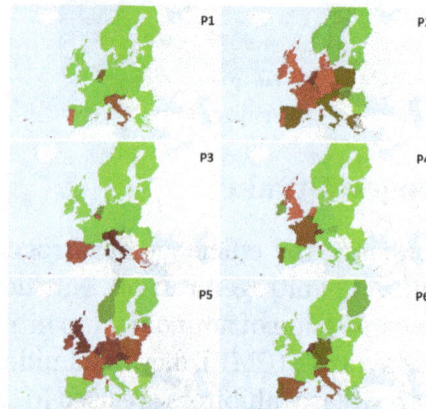

Pesticide categories, EUROSTAT. P5= Plant growth regulators. The red/green scale represents high/low pesticide sales per arable land.

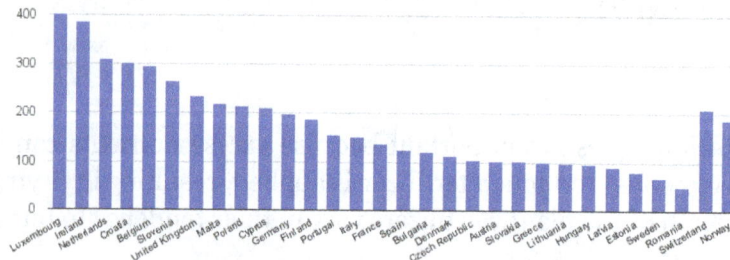

Fertilizer Consumption (kg/ha), The Word Bank 2012

Environmental Effects

Runoff of soil and fertilizer during a rain storm

An algal bloom caused by eutrophication

Water

Agricultural run-off is a major contributor to the eutrophication of fresh water bodies. For example, in the US, about half of all the lakes are eutrophic. The main contributor to eutrophication is phosphate, which is normally a limiting nutrient; high concentrations promote the growth of cyanobacteria and algae, the demise of which consumes oxygen. Cyanobacteria blooms ('algal blooms') can also produce harmful toxins that can accumulate in the food chain, and can be harmful to humans.

The nitrogen-rich compounds found in fertilizer runoff are the primary cause of serious oxygen depletion in many parts of oceans, especially in coastal zones, lakes and rivers. The resulting lack of dissolved oxygen greatly reduces the ability of these areas to sustain oceanic fauna. The number of oceanic dead zones near inhabited coastlines are increasing. As of 2006, the application of nitrogen fertilizer is being increasingly controlled in northwestern Europe and the United States. If eutrophication *can* be reversed, it may take decades before the accumulated nitrates in groundwater can be broken down by natural processes.

Nitrate Pollution

Only a fraction of the nitrogen-based fertilizers is converted to produce and other plant matter. The remainder accumulates in the soil or lost as run-off. High application rates of nitrogen-containing fertilizers combined with the high water solubility of nitrate leads to increased runoff into surface water as well as leaching into groundwater, thereby causing groundwater pollution. The excessive use of nitrogen-containing fertilizers (be they synthetic or natural) is particularly damaging, as much of the nitrogen that is not taken up by plants is transformed into nitrate which is easily leached.

Nitrate levels above 10 mg/L (10 ppm) in groundwater can cause 'blue baby syndrome' (acquired methemoglobinemia). The nutrients, especially nitrates, in fertilizers can cause problems for natural habitats and for human health if they are washed off soil into watercourses or leached through soil into groundwater.

Soil

Acidification

Nitrogen-containing fertilizers can cause soil acidification when added. This may lead to decreases in nutrient availability which may be offset by liming.

Accumulation of Toxic Elements

Cadmium

The concentration of cadmium in phosphorus-containing fertilizers varies considerably and can be problematic. For example, mono-ammonium phosphate fertilizer may have a cadmium content of as low as 0.14 mg/kg or as high as 50.9 mg/kg. This is because the phosphate rock used in their manufacture can contain as much as 188 mg/kg cadmium (examples are deposits on Nauru and the Christmas islands). Continuous use of high-cadmium fertilizer can contaminate soil (as shown in New Zealand) and plants. Limits to the cadmium content of phosphate fertilizers has been considered by the European Commission. Producers of phosphorus-containing fertilizers now select phosphate rock based on the cadmium content.

Fluoride

Phosphate rocks contain high levels of fluoride. Consequently, the widespread use of phosphate fertilizers has increased soil fluoride concentrations. It has been found that food contamination from fertilizer is of little concern as plants accumulate little fluoride from the soil; of greater concern is the possibility of fluoride toxicity to livestock that ingest contaminated soils. Also of possible concern are the effects of fluoride on soil microorganisms.

Radioactive Elements

The radioactive content of the fertilizers varies considerably and depends both on their concentrations in the parent mineral and on the fertilizer production process. Uranium-238 concentrations range can range from 7 to 100 pCi/g in phosphate rock and from 1 to 67 pCi/g in phosphate fertilizers. Where high annual rates of phosphorus fertilizer are used, this can result in uranium-238 concentrations in soils and drainage waters that are several times greater than are normally present. However, the impact of these increases on the risk to human health from radinuclide contamination of foods is very small (less than 0.05 mSv/y).

Other Metals

Steel industry wastes, recycled into fertilizers for their high levels of zinc (essential to plant growth), wastes can include the following toxic metals: lead arsenic, cadmium, chromium, and nickel. The most common toxic elements in this type of fertilizer are mercury, lead, and arsenic. These potentially harmful impurities can be removed; however, this significantly increases cost. Highly pure fertilizers are widely available and perhaps best known as the highly water-soluble fertilizers containing blue dyes used around households, such as Miracle-Gro. These highly water-soluble fertilizers are used in the plant nursery business and are available in larger packages at significantly less cost than retail quantities. There are also some inexpensive retail granular garden fertilizers made with high purity ingredients.

Trace Mineral Depletion

Attention has been addressed to the decreasing concentrations of elements such as iron, zinc, copper and magnesium in many foods over the last 50–60 years. Intensive farming practices, including the use of synthetic fertilizers are frequently suggested as reasons for these declines and organic farming is often suggested as a solution. Although improved crop yields resulting from NPK fertilizers are known to dilute the concentrations of other nutrients in plants, much of the measured decline can be attributed to the use of progressively higher-yielding crop varieties which produce foods with lower mineral concentrations than their less productive ancestors. It is, therefore, unlikely that organic farming or reduced use of fertilizers will solve the problem; foods with high nutrient density are posited to be achieved using older, lower-yielding varieties or the development of new high-yield, nutrient-dense varieties.

Fertilizers are, in fact, more likely to solve trace mineral deficiency problems than cause them: In Western Australia deficiencies of zinc, copper, manganese, iron and molybdenum were identified as limiting the growth of broad-acre crops and pastures in the 1940s and 1950s. Soils in Western Australia are very old, highly weathered and deficient in many of the major nutrients and trace elements. Since this time these trace elements are routinely added to fertilizers used in agriculture in this state. Many other soils around the world are deficient in zinc, leading to deficiency in both plants and humans, and zinc fertilizers are widely used to solve this problem.

Changes in Soil Biology

High levels of fertilizer may cause the breakdown of the symbiotic relationships between plant roots and mycorrhizal fungi.

Energy Consumption and Sustainability

In the USA in 2004, 317 billion cubic feet of natural gas were consumed in the industrial production of ammonia, less than 1.5% of total U.S. annual consumption of natural gas. A 2002 report suggested that the production of ammonia consumes about 5% of global natural gas consumption, which is somewhat under 2% of world energy production.

Ammonia is produced from natural gas and air. The cost of natural gas makes up about 90% of the cost of producing ammonia. The increase in price of natural gases over the past decade, along with other factors such as increasing demand, have contributed to an increase in fertilizer price.

Contribution to Climate Change

The greenhouse gases carbon dioxide, methane and nitrous oxide are produced during the manufacture of nitrogen fertilizer. The effects can be combined into an equivalent amount of carbon dioxide. The amount varies according to the efficiency of the process. The figure for the United Kingdom is over 2 kilogrammes of carbon dioxide equivalent for each kilogramme of ammonium nitrate. Nitrogen fertilizer can be converted by soil bacteria to nitrous oxide, a greenhouse gas.

Atmosphere

Through the increasing use of nitrogen fertilizer, which was used at a rate of about 110 million tons (of N) per year in 2012, adding to the already existing amount of reactive nitrogen, nitrous oxide (N_2O) has become the third most important greenhouse gas after carbon dioxide and methane. It has a global warming potential 296 times larger than an equal mass of carbon dioxide and it also contributes to stratospheric ozone depletion. By changing processes and procedures, it is possible to mitigate some, but not all, of these effects on anthropogenic climate change.

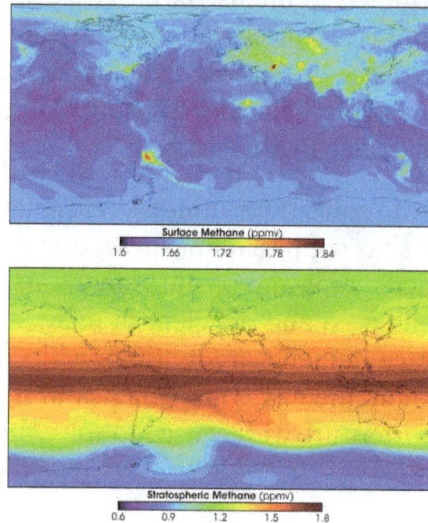

Global methane concentrations (surface and atmospheric) for 2005; note distinct plumes

Methane emissions from crop fields (notably rice paddy fields) are increased by the application of ammonium-based fertilizers. These emissions contribute to global climate change as methane is a potent greenhouse gas.

Regulation

In Europe problems with high nitrate concentrations in run-off are being addressed by the European Union's Nitrates Directive. Within Britain, farmers are encouraged to manage their land more sustainably in 'catchment-sensitive farming'. In the US, high concentrations of nitrate and phosphorus in runoff and drainage water are classified as non-point source pollutants due to their diffuse origin; this pollution is regulated at state level. Oregon and Washington, both in the United States, have fertilizer registration programs with on-line databases listing chemical analyses of fertilizers.

History

Founded in 1812, Mirat, producer of manures and fertilizers, is claimed to be the oldest industrial business in Salamanca (Spain).

Management of soil fertility has been the preoccupation of farmers for thousands of years. Egyptians, Romans, Babylonians, and early Germans all are recorded as using minerals and or manure to enhance the productivity of their farms. The modern science of plant nutrition started in the 19th century and the work of German chemist Justus von Liebig, among others. John Bennet Lawes, an English entrepreneur, began to experiment on the effects of various manures on plants growing in pots in 1837, and a year or two later the experiments were extended to crops in the field. One immediate consequence was that in 1842 he patented a manure formed by treating phosphates with sulphuric acid, and thus was the first to create the artificial manure industry. In the succeeding year he enlisted the services of Joseph Henry Gilbert, with whom he carried on for more than half a century on experiments in raising crops at the Institute of Arable Crops Research.

The Birkeland–Eyde process was one of the competing industrial processes in the beginning of nitrogen based fertilizer production. This process was used to fix atmospheric nitrogen (N_2) into nitric acid (HNO_3), one of several chemical processes generally referred to as nitrogen fixation. The resultant nitric acid was then used as a source of nitrate (NO_3^-). A factory based on the process was built in Rjukan and Notodden in Norway, combined with the building of large hydroelectric power facilities.

The 1910s and 1920s witness the rise of the Haber process and the Ostwald process. The Haber process produces ammonia (NH_3) from methane (CH_4) gas and molecular nitrogen (N_2). The ammonia from the Haber process is then converted into nitric acid (HNO_3) in the Ostwald process. The development of synthetic fertilizer has significantly supported global population growth — it has been estimated that almost half the people on the Earth are currently fed as a result of synthetic nitrogen fertilizer use.

The use of commercial fertilizers has increased steadily in the last 50 years, rising almost 20-fold to the current rate of 100 million tonnes of nitrogen per year. Without commercial fertilizers it is estimated that about one-third of the food produced now could not be produced. The use of phosphate fertilizers has also increased from 9 million tonnes per year in 1960 to 40 million tonnes per year in 2000. A maize crop yielding 6–9 tonnes of grain per hectare (2.5 acres) requires 31–50 kilograms (68–110 lb) of phosphate fertilizer to be applied; soybean crops require about half, as 20–25 kg per hectare. Yara International is the world's largest producer of nitrogen-based fertilizers.

Controlled-nitrogen-release technologies based on polymers derived from combining urea and formaldehyde were first produced in 1936 and commercialized in 1955. The early product had 60 percent of the total nitrogen cold-water-insoluble, and the unreacted (quick-release) less than 15%. Methylene ureas were commercialized in the 1960s and 1970s, having 25% and 60% of the nitrogen as cold-water-insoluble, and unreacted urea nitrogen in the range of 15% to 30%.

In the 1960s, the Tennessee Valley Authority National Fertilizer Development Center began developing sulfur-coated urea; sulfur was used as the principal coating material because of its low cost and its value as a secondary nutrient. Usually there is another wax or polymer which seals the sulfur; the slow-release properties depend on the degradation of the secondary sealant by soil microbes as well as mechanical imperfections (cracks, etc.) in the sulfur. They typically provide 6 to 16 weeks of delayed release in turf applications. When a hard polymer is used as the secondary coating, the properties are a cross between diffusion-controlled particles and traditional sulfur-coated.

Plant Nutrients

Chemical elements that are essential for the proper development and growth of plants are typically referred to as plant nutrients. The list of plant nutrients recognized as being necessary for plant growth has increased over the years and now totals sixteen.

Plant nutrition is the study of the chemical elements and compounds necessary for plant growth, plant metabolism and their external supply. In 1972, Emanuel Epstein defined two criteria for an element to be essential for plant growth:

1. in its absence the plant is unable to complete a normal life cycle.

2. or that the element is part of some essential plant constituent or metabolite.

This is in accordance with Justus von Liebig's law of the minimum. The essential plant nutrients include carbon, oxygen and hydrogen which are absorbed from the air, whereas other nutrients including nitrogen are typically obtained from the soil (exceptions include some parasitic or carnivorous plants).

Plants must obtain the following mineral nutrients from their growing medium:

- the macronutrients: nitrogen (N), phosphorus (P), potassium (K), calcium (Ca), sulfur (S), magnesium (Mg), sodium (Na)

- the micronutrients (or trace minerals): boron (B), chlorine (Cl), manganese (Mn), iron (Fe), zinc (Zn), copper (Cu), molybdenum (Mo), nickel (Ni). and cobalt (Co)

The macronutrients are consumed in larger quantities; hydrogen, oxygen, nitrogen and carbon contribute to over 95% of a plants' entire biomass on a dry matter weight basis. Micronutrients are present in plant tissue in quantities measured in parts per million, ranging from 0.1 to 200 ppm, or less than 0.02% dry weight.

Most soil conditions across the world can provide plants adapted to that climate and soil with sufficient nutrition for a complete life cycle, without the addition of nutrients as fertilizer. However, if the soil is cropped it is necessary to artificially modify soil fertility through the addition of fertilizer to promote vigorous growth and increase or sustain yield. This is done because, even with adequate water and light, nutrient deficiency can limit growth and crop yield.

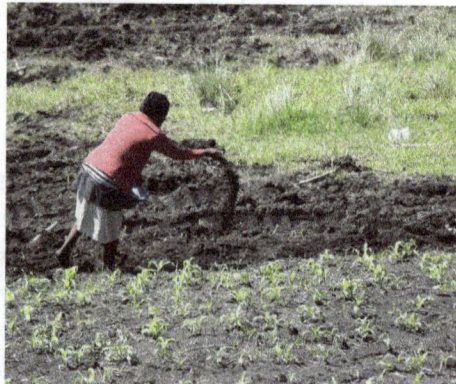

Farmer spreading decomposing manure to improve soil fertility and plant nutrition

Processes

Plants take up essential elements from the soil through their roots and from the air (mainly consisting of nitrogen and oxygen) through their leaves. Nutrient uptake in the soil is achieved by cation exchange, wherein root hairs pump hydrogen ions (H^+) into the soil through proton pumps. These hydrogen ions displace cations attached to negatively charged soil particles so that the cations are available for uptake by the root. In the leaves, stomata open to take in carbon dioxide and expel oxygen. The carbon dioxide molecules are used as the carbon source in photosynthesis.

The root, especially the root hair, is the essential organ for the uptake of nutrients. The structure and architecture of the root can alter the rate of nutrient uptake. Nutrient ions are transported to the center of the root, the stele, in order for the nutrients to reach the conducting tissues, xylem and phloem. The Casparian strip, a cell wall outside the stele but within the root, prevents passive flow of water and nutrients, helping to regulate the uptake of nutrients and water. Xylem moves water and mineral ions within the plant and phloem accounts for organic molecule transportation. Water potential plays a key role in a plant's nutrient uptake. If the water potential is more negative within the plant than the surrounding soils, the nutrients will move from the region of higher solute concentration—in the soil—to the area of lower solute concentration - in the plant.

There are three fundamental ways plants uptake nutrients through the root:

1. Simple diffusion occurs when a nonpolar molecule, such as O_2, CO_2, and NH_3 follows a concentration gradient, moving passively through the cell lipid bilayer membrane without the use of transport proteins.

2. Facilitated diffusion is the rapid movement of solutes or ions following a concentration gradient, facilitated by transport proteins.

3. Active transport is the uptake by cells of ions or molecules against a concentration gradient; this requires an energy source, usually ATP, to power molecular pumps that move the ions or molecules through the membrane.

Nutrients can be moved within plants to where they are most needed. For example, a plant will try to supply more nutrients to its younger leaves than to its older ones. When nutrients are mobile within the plant, symptoms of any deficiency become apparent first on the older leaves. However, not all nutrients are equally mobile. Nitrogen, phosphorus, and potassium are mobile nutrients while the others have varying degrees of mobility. When a less-mobile nutrient is deficient, the younger leaves suffer because the nutrient does not move up to them but stays in the older leaves. This phenomenon is helpful in determining which nutrients a plant may be lacking.

Many plants engage in symbiosis with microorganisms. Two important types of these relationship are

1. with bacteria such as rhizobia, that carry out biological nitrogen fixation, in which atmospheric nitrogen (N_2) is converted into ammonium (NH_4^+); and

2. with mycorrhizal fungi, which through their association with the plant roots help to create a larger effective root surface area. Both of these mutualistic relationships enhance nutrient uptake.

Though nitrogen is plentiful in the Earth's atmosphere, relatively few plants harbour nitrogen-fixing bacteria, so most plants rely on nitrogen compounds present in the soil to support their growth. These can be supplied by mineralization of soil organic matter or added plant residues, nitrogen fixing bacteria, animal waste, through the breaking of triple bonded N_2 molecules by lightening strikes or through the application of fertilizers.

Functions of Nutrients

At least 17 elements are known to be essential nutrients for plants. In relatively large amounts, the soil supplies nitrogen, phosphorus, potassium, calcium, magnesium, and sulfur; these are often called the macronutrients. In relatively small amounts, the soil supplies iron, manganese, boron, molybdenum, copper, zinc, chlorine, and cobalt, the so-called micronutrients. Nutrients must be available not only in sufficient amounts but also in appropriate ratios.

Plant nutrition is a difficult subject to understand completely, partially because of the variation between different plants and even between different species or individuals of a given clone. Elements present at low levels may cause deficiency symptoms, and toxicity is possible at levels that are too high. Furthermore, deficiency of one element may present as symptoms of toxicity from another element, and vice versa. An abundance of one nutrient may cause a deficiency of another nutrient. For example, K^+ uptake can be influenced by the amount of NH_4^+ available.

Although nitrogen is plentiful in the Earth's atmosphere, relatively few plants engage in nitrogen fixation (conversion of atmospheric nitrogen to a biologically useful form). Most plants, therefore, require nitrogen compounds to be present in the soil in which they grow.

Carbon and oxygen are absorbed from the air while other nutrients are absorbed from the soil. Green plants obtain their carbohydrate supply from the carbon dioxide in the air by the process of photosynthesis. Each of these nutrients is used in a different place for a different essential function.

Macronutrients (Derived from Air and Water)

Carbon

Carbon forms the backbone of most plant biomolecules, including proteins, starches and cellulose. Carbon is fixed through photosynthesis; this converts carbon dioxide from the air into carbohydrates which are used to store and transport energy within the plant.

Hydrogen

Hydrogen also is necessary for building sugars and building the plant. It is obtained almost entirely from water. Hydrogen ions are imperative for a proton gradient to help drive the electron transport chain in photosynthesis and for respiration.

Oxygen

Oxygen is a component of many organic and inorganic molecules within the plant, and is acquired in many forms. These include: O_2 and CO_2 (mainly from the air via leaves) and H_2O, NO_3^-, $H_2PO_4^-$ and $SO2_4^-$ (mainly from the soil water via roots). Plants produce oxygen gas (O_2) along with glucose during photosynthesis but then require O_2 to undergo aerobic cellular respiration and break down this glucose to produce ATP.

Macronutrients (Primary)

Nitrogen

Nitrogen is a major constituent of several of the most important plant substances. For example, nitrogen compounds comprise 40% to 50% of the dry matter of protoplasm, and it is a constituent of amino acids, the building blocks of proteins. It is also an essential constituent of chlorophyll. Nitrogen deficiency most often results in stunted growth, slow growth, and chlorosis. Nitrogen deficient plants will also exhibit a purple appearance on the stems, petioles and underside of leaves from an accumulation of anthocyanin pigments. Most of the nitrogen taken up by plants is from the soil in the forms of NO_3^-, although in acid environments such as boreal forests where nitrification is less likely to occur, ammonium NH_4^+ is more likely to be the dominating source of nitrogen. Amino acids and proteins can only be built from NH_4^+, so NO_3^- must be reduced. In many agricultural settings, nitrogen is the limiting nutrient for rapid growth. Nitrogen is transported via the xylem from the roots to the leaf canopy as nitrate ions, or in an organic form, such as amino acids or amides. Nitrogen can also be transported in the phloem sap as amides, amino acids and ureides; it is therefore mobile within the plant, and the older leaves exhibit chlorosis and necrosis earlier than the younger leaves.

There is an abundant supply of nitrogen in the earth's atmosphere — N_2 gas comprises nearly 79% of air. However, N_2 is unavailable for use by most organisms because there is a triple bond between the two nitrogen atoms in the molecule, making it almost inert. In order for nitrogen to be used for growth it must be "fixed" (combined) in the form of ammonium (NH_4) or nitrate (NO_3) ions. The weathering of rocks releases these ions so slowly that it has a negligible effect on the availability of fixed nitrogen. Therefore, nitrogen is often the limiting factor for growth and biomass production in all environments where there is a suitable climate and availability of water to support life.

Nitrogen enters the plant largely through the roots. A "pool" of soluble nitrogen accumulates. Its composition within a species varies widely depending on several factors, including day length, time of day, night temperatures, nutrient deficiencies, and nutrient imbalance. Short day length promotes asparagine formation, whereas glutamine is produced under long day regimes. Darkness favors protein breakdown accompanied by high asparagine accumulation. Night temperature modifies the effects due to night length, and soluble nitrogen tends to accumulate owing to retarded synthesis and breakdown of proteins. Low night temperature conserves glutamine; high night temperature increases accumulation of asparagine because of breakdown. Deficiency of K accentuates differences between long- and short-day plants. The pool of soluble nitrogen is much smaller than in well-nourished plants when N and P are deficient since uptake of nitrate and further reduction and conversion of N to organic forms is restricted more than is protein synthesis. Deficiencies of Ca, K, and S affect the conversion of organic N to protein more than uptake and reduction. The size of the pool of soluble N is no guide *per se* to growth rate, but the size of the pool in relation to total N might be a useful ratio in this regard. Nitrogen availability in the rooting medium also affects the size and structure of tracheids formed in the long lateral roots of white spruce (Krasowski and Owens 1999).

Microorganisms have a central role in almost all aspects of nitrogen availability, and therefore for life support on earth. Some bacteria can convert N_2 into ammonia by the process termed *nitrogen fixation*; these bacteria are either free-living or form symbiotic associations with plants or other organisms (e.g., termites, protozoa), while other bacteria bring about transformations of ammonia to nitrate, and of nitrate to N_2 or other nitrogen gases. Many bacteria and fungi degrade organic matter, releasing fixed nitrogen for reuse by other organisms. All these processes contribute to the nitrogen cycle.

Phosphorus

Like nitrogen, phosphorus is involved with many vital plant processes. Within a plant, it is present mainly as a structural component of the nucleic acids: deoxyribonucleic acid (DNA) and ribonucleic acid (RNA), as well as a constituent of fatty phospholipids, that are important in membrane development and function. It is present in both organic and inorganic forms, both of which are readily translocated within the plant. All energy transfers in the cell are critically dependent on phosphorus. As with all living things, phosphorus is part of the Adenosine triphosphate (ATP), which is of immediate use in all processes that require energy with the cells. Phosphorus can also be used to modify the activity of various enzymes by phosphorylation, and is used for cell signaling. Phosphorus is concentrated at the most actively growing points of a plant and stored within seeds in anticipation of their germination. Phosphorus is most commonly found in the soil in the form of polyprotic phosphoric acid (H_3PO_4), but is taken up most readily in the form of $H_2PO_4^-$. Phosphorus is available to plants in limited quantities in most soils because it is released very slowly from insoluble phosphates and is rapidly fixed once again. Under most environmental conditions it is the element that limits growth because of this constriction and due to its high demand by plants and microorganisms. Plants can increase phosphorus uptake by a mutualism with mycorrhiza. A Phosphorus deficiency in plants is characterized by an intense green coloration or reddening in leaves due to lack of chlorophyll. If the plant is experiencing high phosphorus deficiencies the leaves may become denatured and show signs of death. Occasionally the leaves may appear purple from an accumulation of anthocyanin. Because phosphorus is a mobile nutrient, older leaves will show the first signs of deficiency.

On some soils, the phosphorus nutrition of some conifers, including the spruces, depends on the ability of mycorrhizae to take up, and make soil phosphorus available to the tree, hitherto unobtainable to the non-mycorrhizal root. Seedling white spruce, greenhouse-grown in sand testing negative for phosphorus, were very small and purple for many months until spontaneous mycorrhizal inoculation, the effect of which was manifested by a greening of foliage and the development of vigorous shoot growth.

Phosphorus deficiency can produce symptoms similar to those of nitrogen deficiency, but as noted by Russel: "Phosphate deficiency differs from nitrogen deficiency in being extremely difficult to diagnose, and crops can be suffering from extreme starvation without there being any obvious signs that lack of phosphate is the cause". Russell's observation applies to at least some coniferous seedlings, but Benzian found that although response to phosphorus in very acid forest tree nurseries in England was consistently high, no species (including Sitka spruce) showed any visible symptom of deficiency other than a slight lack of lustre. Phosphorus levels have to be exceedingly low before visible symptoms appear in such seedlings. In sand culture at 0 ppm phosphorus, white spruce seedlings were very small and tinted deep purple; at 0.62 ppm, only the smallest seedlings were deep purple; at 6.2 ppm, the seedlings were of good size and color.

It is useful to apply a high phosphorus content fertilizer, such as bone meal, to perennials to help with successful root formation.

Potassium

Unlike other major elements, potassium does not enter into the composition of any of the important plant constituents involved in metabolism, but it does occur in all parts of plants in substantial amounts. It seems to be of particular importance in leaves and at growing points. Potassium is outstanding among the nutrient elements for its mobility and solubility within plant tissues. Processes involving potassium include the formation of carbohydrates and proteins, the regulation of internal plant moisture, as a catalyst and condensing agent of complex substances, as an accelerator of enzyme action, and as contributor to photosynthesis, especially under low light intensity.

Potassium regulates the opening and closing of the stomata by a potassium ion pump. Since stomata are important in water regulation, potassium regulates water loss from the leaves and increases drought tolerance. Potassium deficiency may cause necrosis or interveinal chlorosis. The potassium ion (K^+) is highly mobile and can aid in balancing the anion (negative) charges within the plant. Potassium helps in fruit coloration, shape and also increases its brix. Hence, quality fruits are produced in potassium-rich soils. Potassium serves as an activator of enzymes used in photosynthesis and respiration. Potassium is used to build cellulose and aids in photosynthesis by the formation of a chlorophyll precursor. Potassium deficiency may result in higher risk of pathogens, wilting, chlorosis, brown spotting, and higher chances of damage from frost and heat.

When soil-potassium levels are high, plants take up more potassium than needed for healthy growth. The term *luxury consumption* has been applied to this. When potassium is moderately deficient, the effects first appear in the older tissues, and from there progress towards the growing points. Acute deficiency severely affects growing points, and die-back commonly occurs. Symptoms of potassium deficiency in white spruce include: browning and death of needles (chlorosis); reduced growth in height and diameter; impaired retention of needles; and reduced needle length. A relationship between potassium nutrition and cold resistance has been found in several tree species, including two species of spruce.

Macronutrients (Secondary and Tertiary)

Sulfur

Sulfur is a structural component of some amino acids and vitamins, and is essential in the manufacturing of chloroplasts. Sulphur is also found in the iron-sulphur complexes of the electron transport chains in photosynthesis. Sulphate ions are mobile and its deficiency, therefore, affects older tissues first. Symptoms of deficiency include yellowing of leaves and stunted growth.

Calcium

Calcium regulates transport of other nutrients into the plant and is also involved in the activation of certain plant enzymes. Calcium deficiency results in stunting. This nutrient is involved in photosynthesis and plant structure. Blossom end rot is also a result of inadequate calcium.

Another common symptom of calcium deficiency in leaves is the curling of the leaf towards the veins or center of the leaf. Many times this can also have a blackened appearance Calcium has been found to have a positive effect in combating salinity in soils. It has been shown to ameliorate the negative effects that salinity has such as reduced water usage of plants. Calcium in plants occurs chiefly in the leaves, with lower concentrations in seeds, fruits, and roots. A major function is as a constituent of cell walls. When coupled with certain acidic compounds of the jelly-like pectins of the middle lamella, calcium forms an insoluble salt. It is also intimately involved in meristems, and is particularly important in root development, with roles in cell division, cell elongation, and the detoxification of hydrogen ions. Other functions attributed to calcium are; the neutralization of organic acids; inhibition of some potassium-activated ions; and a role in nitrogen absorption. A notable feature of calcium-deficient plants is a defective root system. Roots are usually affected before above-ground parts.

Magnesium

The outstanding role of magnesium in plant nutrition is as a constituent of the chlorophyll molecule. As a carrier, it is also involved in numerous enzyme reactions as an effective activator, in which it is closely associated with energy-supplying phosphorus compounds. Magnesium is very mobile in plants, and, like potassium, when deficient is translocated from older to younger tissues, so that signs of deficiency appear first on the oldest tissues and then spread progressively to younger tissues.

Micro-nutrients

Plants are able sufficiently to accumulate most trace elements. Some plants are sensitive indicators of the chemical environment in which they grow (Dunn 1991), and some plants have barrier mechanisms that exclude or limit the uptake of a particular element or ion species, e.g., alder twigs commonly accumulate molybdenum but not arsenic, whereas the reverse is true of spruce bark (Dunn 1991). Otherwise, a plant can integrate the geochemical signature of the soil mass permeated by its root system together with the contained groundwaters. Sampling is facilitated by the tendency of many elements to accumulate in tissues at the plant's extremities.

Iron

Iron is necessary for photosynthesis and is present as an enzyme cofactor in plants. Iron deficiency can result in interveinal chlorosis and necrosis. Iron is not a structural part of chlorophyll but very much essential for its synthesis. Copper deficiency can be responsible for promoting an iron deficiency. It helps in the electron transport of plant.

Molybdenum

Molybdenum is a cofactor to enzymes important in building amino acids and is involved in nitrogen metabolism. Molybdenum is part of the nitrate reductase enzyme (needed for the reduction of nitrate) and the nitrogenase enzyme (required for biological nitrogen fixation).

Boron

Boron is found in the highly insoluble mineral, tourmaline (a crystalline boron silicate mineral). It is absorbed by plants in the form of the anion BO_3^{3}. It is available to plants in moderately soluble mineral forms of Ca, Mg and Na borates and the highly soluble form of organic compounds. Concentration in soil must, in general, be below 5 ppm in the soil water solution, above that toxicity results. Its availability in soils ranges from 20 to 200 pounds per acre in the first eight inches, of which less than 5% is available. It is available to plants over a range of pH, from 5.0 to 7.5. It is mobile in the soil, hence, it is prone to leaching. Leaching removes substantial amounts of boron in sandy soil, but little in fine silt or clay soil. Boron's fixation to those minerals at high pH can render boron unavailable, while low pH frees the fixed boron, leaving it prone to leaching in wet climates. It precipitates with other minerals in the form of borax in which form it was first used over 400 years ago as a soil supplement. Decomposition of organic material causes boron to be deposited in the topmost soil layer; organic forms of boron are more soluble than their mineral form, hence are more available in the top few inches. When soil dries it can cause a precipitous drop in the availability of boron to plants as the plants cannot draw nutrients from that desiccated layer. Hence, boron deficiency diseases appear in dry weather.

Boron has many functions within a plant: it affects flowering and fruiting, pollen germination, cell division, and active salt absorption. The metabolism of amino acids and proteins, carbohydrates, calcium, and water are strongly affected by boron. Many of those listed functions may be embodied by its function in moving the highly polar sugars through cell membranes by reducing their polarity and hence the energy needed to pass the sugar. If sugar cannot pass to the fastest growing parts rapidly enough, those parts die. Boron is relatively immobile within a plant suggesting that the molecule is fixed to the points in the membrane where they facilitate sugar transport.

Boron is not relocatable in the plant via the phloem. It must be supplied to the growing parts via the xylem. Foliar sprays affect only those parts sprayed, which may be insufficient for the fastest growing parts, and is very temporary.

Boron is essential for the proper forming and strengthening of cell walls. Lack of boron results in short thick cells producing stunted fruiting bodies and roots. Calcium to boron ratio must be maintained in a narrow range for normal plant growth. For alfalfa, that calcium to boron ratio must be from 80:1 to 600:1. Boron deficiency appears at 800:1 and higher. For alfalfa, similar ratios exist for magnesium, copper, nitrogen and potassium. Boron levels within plants differ with plant species and range from 2.3 p.p.m for barley to 94.7 p.p.m for poppy . Lack of boron causes failure of calcium metabolism which produces hollow heart in beets and peanuts.

Inadequate amounts of boron affect many agricultural crops, legume forage crops most strongly. Of the micronutrients, boron deficiencies are second most common after zinc. Deficiencies of boron when soil is cropped are common and require the application of mineral supplement; one ton of alfalfa hay carries with it one ounce of boron, 100 bushels of peaches 4 ounces. Deficiency results in the death of the terminal growing points. Symptoms first appear as stunted growth, then to cellular changes, which leads to physical changes, and finally death of the plant.

Boron supplements derive from dry lake bed deposits such as those in Death Valley, USA, in the form of sodium tetraborate (borax), from which less soluble calcium borate is made. Foliar sprays are used on fruit crop trees in soils of high alkalinity. Boron is often applied to fields as a contaminant in other soil amendments but is not generally adequate to make up the rate of loss by cropping. The rates of application of borate to produce an adequate alfalfa crop range from 15 pounds per acre for a sandy-silt, acidic soil of low organic matter, to 60 pounds per acre for a soil with high organic matter, high cation exchange capacity and high pH.

Boron concentration in soil water solution higher than one ppm is toxic to most plants. Toxic concentrations within plants are 10 to 50 ppm for small grains and 200 ppm in boron-tolerant crops such as sugar beets, rutabaga, cucumbers, and conifers. Toxic soil conditions are generally limited to arid regions or can be caused by underground borax deposits in contact with water or volcanic gases dissolved in percolating water. Application rates should be limited to a few pounds per acre in a test plot to determine if boron is needed generally. Otherwise, testing for boron levels in plant material is required to determine remedies. Excess boron can be removed by irrigation and assisted by application of elemental sulfur to lower the pH and increase boron's solubility. Application of calcium will increase soil alkalinity, causing boron to fix on the mineral soil component and remove some fraction, thereby reducing boron toxicity.

Boron deficiencies must be detected by analysis of plant material to apply a correction before the obvious symptoms appear, after which it is too late to prevent crop loss. Strawberries deficient in boron will produce lumpy fruit; apricots will not blossom or, if they do, will not fruit or will drop their fruit depending on the level of boron deficit. Broadcast of boron supplements is effective and long term; a foliar spray is immediate but must be repeated.

Boron is an essential element for the health of animals which derive their boron from plant material.

Copper

Copper is important for photosynthesis. Symptoms for copper deficiency include chlorosis.It is involved in many enzyme processes; necessary for proper photosynthesis; involved in the manufacture of lignin (cell walls) and involved in grain production. It is also hard to find in some soil conditions.

Manganese

Manganese is necessary for photosynthesis, including the building of chloroplasts. Manganese deficiency may result in coloration abnormalities, such as discolored spots on the foliage.

Sodium

Sodium is involved in the regeneration of phosphoenolpyruvate in CAM and C_4 plants. Sodium can potentially replace potassium's regulation of stomatal opening and closing.

Essentiality of sodium:

- Essential for C_4 plants rather C_3

- Substitution of K by Na: Plants can be classified into four groups:

  o Group A—a high proportion of K can be replaced by Na and stimulate the growth, which cannot be achieved by the application of K

  o Group B—specific growth responses to Na are observed but they are much less distinct

  o Group C—Only minor substitution is possible and Na has no effect

  o Group D—No substitution occurs

- Stimulate the growth—increase leaf area and stomata. Improves the water balance

- Na functions in metabolism

  o C4 metabolism

  o Impair the conversion of pyruvate to phosphoenol-pyruvate

  o Reduce the photosystem II activity and ultrastructural changes in mesophyll chloroplast

- Replacing K functions

  o Internal osmoticum

  o Stomatal function

  o Photosynthesis

  o Counteraction in long distance transport

  o Enzyme activation

- Improves the crop quality e.g. improves the taste of carrots by increasing sucrose

Zinc

Zinc is required in a large number of enzymes and plays an essential role in DNA transcription. A typical symptom of zinc deficiency is the stunted growth of leaves, commonly known as "little leaf" and is caused by the oxidative degradation of the growth hormone auxin.

Nickel

In higher plants, nickel is absorbed by plants in the form of Ni^{2+} ion. Nickel is essential for activation of urease, an enzyme involved with nitrogen metabolism that is required to process urea. Without nickel, toxic levels of urea accumulate, leading to the formation of necrotic lesions. In lower plants, nickel activates several enzymes involved in a variety of processes, and can substitute for zinc and iron as a cofactor in some enzymes.

Chlorine

Chlorine, as compounded chloride, is necessary for osmosis and ionic balance; it also plays a role in photosynthesis.

Cobalt

Cobalt has proven to be beneficial to at least some plants although it does not appear to be essential for most species. It has, however, been shown to be essential for nitrogen fixation by the nitrogen-fixing bacteria associated with legumes and other plants.

Aluminum

Aluminum is one of the few elements capable of making soil more acidic. This is achieved by aluminum taking hydroxide ions out of water, leaving hydrogen ions behind. As a result, the soil is more acidic, which makes it unlivable for many plants. Another consequence of aluminum in soils is aluminum toxicity, which inhibits root growth.

- Tea has a high tolerance for aluminum (Al) toxicity and the growth is stimulated by Al application. The possible reason is the prevention of Cu, Mn or P toxicity effects.

- There have been reports that Al may serve as a fungicide against certain types of root rot.

Silicon

Silicon is not considered an essential element for plant growth and development. It is always found in abundance in the environment and hence if needed it is available. It is found in the structures of plants and improves the health of plants.

In plants, silicon has been shown in experiments to strengthen cell walls, improve plant strength, health, and productivity. There have been studies showing evidence of silicon improving drought and frost resistance, decreasing lodging potential and boosting the plant's natural pest and disease fighting systems. Silicon has also been shown to improve plant vigor and physiology by improving root mass and density, and increasing above ground plant biomass and crop yields. Silicon is currently under consideration by the Association of American Plant Food Control Officials (AAPFCO) for elevation to the status of a "plant beneficial substance".

Vanadium

Vanadium may be required by some plants, but at very low concentrations. It may also be substituting for molybdenum.

Selenium

Selenium is probably not essential for flowering plants, but it can be beneficial; it can stimulate plant growth, improve tolerance of oxidative stress, and increase resistance to pathogens and herbivory.

Selenium is, however, an essential mineral element for animal (including human) nutrition and selenium deficiencies are known to occur when food or animal feed is grown on selenium-deficient soils. The use of inorganic selenium fertilizers can increase selenium concentrations in edible crops and animal diets thereby improving animal health.

Nutrient Deficiency

The effect of a nutrient deficiency can vary from a subtle depression of growth rate to obvious stunting, deformity, discoloration, distress, and even death. Visual symptoms distinctive enough to be useful in identifying a deficiency are rare. Most deficiencies are multiple and moderate. However, while a deficiency is seldom that of a single nutrient, nitrogen is commonly the nutrient in shortest supply.

Chlorosis of foliage is not always due to mineral nutrient deficiency. Solarization can produce superficially similar effects, though mineral deficiency tends to cause premature defoliation, whereas solarization does not, nor does solarization depress nitrogen concentration.

Nutrient Status of Plants

Nutrient status (mineral nutrient and trace element composition, also called ionome and nutrient profile) of plants are commonly portrayed by tissue elementary analysis. Interpretation of the results of such studies, however, has been controversial. During the last decades the nearly two-century-old "law of minimum" or "Liebig's law" (that states that plant growth is controlled not by the total amount of resources available, but by the scarcest resource) has been replaced by several mathematical approaches that use different models in order to take the interactions between the individual nutrients into account. The latest developments in this field are based on the fact that the nutrient elements (and compounds) do not act independently from each other; Baxter, 2015, because there may be direct chemical interactions between them or they may influence each other's uptake, translocation, and biological action via a number of mechanisms as exemplified for the case of ammonia.

Plant Nutrition in Agricultural Systems

Hydroponics

Hydroponics is a method for growing plants in a water-nutrient solution without the use of nutrient-rich soil. It allows researchers and home gardeners to grow their plants in a controlled environment. The most common solution is the Hoagland solution, developed by D. R. Hoagland in 1933. The solution consists of all the essential nutrients in the correct proportions necessary for most plant growth. An aerator is used to prevent an anoxic event or hypoxia. Hypoxia can affect nutrient uptake of a plant because, without oxygen present, respiration becomes inhibited within the root cells. The nutrient film technique is a hydroponic technique in which the roots are not fully submerged. This allows for adequate aeration of the roots, while a "film" thin layer of nutrient-rich water is pumped through the system to provide nutrients and water to the plant.

Expression

Many countries express quantities or percentages of the primary nutrients in terms of elemental nitrogen (N), phosphorus pen oxide (P_2O_5), and potassium oxide (K_2O_2). Secondary nutrients and micronutrients usually are expressed on an elemental basis although calcium and magnesium sometimes are expressed in the oxide form. However, several countries express all plant nutrients on an elemental basis.

Classification of Elements Essential for Plant Growth		
Major elements (Macronutrients)	(Available from air or water)	Carbon
		Hydrogen, Oxygen
	Primary nutrients	Nitrogen, Phosphorus, Potassium
	Secondary nutrients	Calcium, Magnesium, Sulfur
Minor elements (Micronutrients)		Boron, Chlorine, Copper Iron Manganese, Molybdenum Zinc

Fertilizer Grade

All fertilizer labels have three bold numbers. The first number is the amount of nitrogen (N), the second number is the amount of phosphate (P_2O_5) and the third number is the amount of potash (K_2O). These three numbers represent the primary nutrients (nitrogen(N) - phosphorus(P) - potassium(K)).

This label, known as the fertilizer grade, is a national standard. A bag of 10-10-10 fertilizer contains 10 percent nitrogen, 10 percent phosphate and 10 percent potash.

Fertilizer grades are made by mixing two or more nutrient sources together to form a blend that is why they are called "mixed fertilizers." Blends contain particles of more than one color. Manufacturers produce different grades for the many types of plants.

You can also get fertilizers that contain only one of each of the primary nutrients. Nitrogen sources include ammonium nitrate (33.5-0-0), urea nitrogen (46-0-0), sodium nitrate (16-0-0) and liquid nitrogen (30-0-0). Phosphorus is provided as 0-46-0 and potash as 0-0-60 or 0-0-50.

Labeling of Fertilizer

The labeling of fertilizers varies by country in terms of analysis methodology, nutrient labeling, and minimum nutrient requirements. The most common labeling convention shows the amounts of nitrogen, phosphorus, and potassium in the fertilizer.

Labeling of Macronutrient Fertilizers

Macronutrient fertilizers are generally labeled with an *NPK* analysis, based on the relative content of the chemical elements nitrogen (N), phosphorus (P), and potassium (K) that are commonly used in fertilizers. However, numbers used in this labeling scheme do not directly represent the source composition or absolute nutrient content of the fertilizer. The N value is the percentage of elemental nitrogen by weight in the fertilizer. The value for P is the fraction by weight of P_2O_5 in a fertilizer with the same amount of phosphorus that gets all of its phosphorus from P_2O_5. The value for K is analogous, based on a fertilizer with K_2O.

For example, the fertilizer *sylvite* is a naturally occurring mineral consisting mostly of potassium chloride (*KCl*). As such, it contains one potassium atom for every chlorine atom, and is 52% potassium and 48% chlorine *by weight* (because of the difference in atomic weights of the elements). K_2O is similarly 83% potassium. Therefore, a fertilizer that gets all its potassium from KCl would have to be 63% K_2O (.52/.83 is .63). Pure KCl fertilizer would thus be labeled 0-0-63; because sylvite is less than pure (it contains other compounds that contain no potassium), it is labeled 0-0-60.

Converting Nutrient Analysis to Composition

The factors for converting from P_2O_5 and K_2O values on a fertilizer label to the concentrations (by weight) of P and K elements are as follows:

- P_2O_5 consists of 56.4% oxygen and 43.6% elemental phosphorus. The percentage (mass fraction) of elemental phosphorus is 43.6% so elemental P = 0.436 x P_2O_5

- K_2O consists of 17% oxygen and 83% elemental potassium. The percentage (mass fraction) of elemental potassium is 83% so elemental K = 0.83 x K_2O

- Nitrogen values represent actual nitrogen content so these numbers do not need to be converted.

Using these conversion factors, an 18–51–20 fertilizer contains by weight:

- 18% elemental (N)

- 22% elemental (P), and

- 17% elemental (K)

Other Labeling Conventions

In the U.K., fertilizer labeling regulations allow for reporting the elemental mass fractions of phosphorus and potassium. The regulations stipulate that this should be done in parentheses after the standard N-P-K values. In Australia, macronutrient fertilizers are labeled with an "N-P-K-S" system, which uses elemental mass fractions rather than standard N-P-K values and includes the amount of sulfur (S) contained in the fertilizer.

NPK Values for Various Synthetic Fertilizers

- 15-00-00 Calcium nitrate

- 21-00-00 Ammonium sulfate

- 30-00-00 to 40-00-00 Sulfur-coated urea (slow release)

- 31-00-00 Isobutylidene diurea (~90% slow release)

- 33-00-00 to 34-00-00 Ammonium nitrate

- 35-00-00 Ureaform (~85% slow release, sparingly soluble ureaformaldehyde)
- 40-00-00 Methylene ureas (~70% slow release)
- 46-00-00 Urea (U-46)
- 82-00-00 Anhydrous ammonia
- 10-34-00 to 11-37-00 Ammonium polyphosphate
- 11-48-00 to 11-55-00 Monoammonium phosphate
- 18-46-00 to 21-54-00 Diammonium phosphate
- 13-00-44 Potassium nitrate
- 00-17-00 to 00-22-00 Superphosphate (Monocalcium phosphate monohydrate with gypsum)
- 00-44-00 to 00-52-00 Triple superphosphate (Monocalcium phosphate monohydrate)

NPK Values for Mined Fertilizer Minerals

- 11-08-02 to 16-12-03 bird guano
- 00-3-00 to 00-8-00 Raw Phosphate Rock (would be 00-34-00 if it were soluble)
- 00-00-22 Potassium magnesium sulfate (K-mag)
- 00-00-60 Potassium chloride

NPK Values for Biosolids Fertilizers and Others

- 09-00-00 dairy manure
- 01-00-01 horse manure
- 03-02-02 poultry manure
- 04-12-00 Bone meal
- 05-05-06 Fish blood and bone
- 06-02-00 Milorganite

Fertilizer Specifications

Specifications are the requirements with which a fertilizer should conform, as agreed upon between buyer and seller. Fertilizer specifications meet differing requirements depending on the use or intent of the specification information.

Specifications are normally used in the contract between the buyer and seller of a fertilizer to ensure agreement on product characteristics or more often to define the product in sufficient detail to effect the satisfaction of both buyer and seller.

Terminology and Definitions

The below specified definitions are those given by International Association for Standardization (ISO) and Association of American Plant Food Control Officials (AAPFCO)

Fertilizer Material- A fertilizer that meets any of the following conditions (AAPFCO):

1. Contains important quantities of no more than one of the primary plant nutrients (nitrogen, phosphorus, or potassium).

2. Has 85% or more of its plant nutrient content present in the form of a single chemical compound.

3. Is derived from a plant or animal residue or by product or natural material deposit which has been processed in such a way that its content of plant nutrients has not been materially changed except by purification and concentration.

Fertilizer- In the simplest terminology, a material, the main function of which is to provide plant nutrients.

Soil Conditioner – Material added to soils, the main function of which is to improve their physical and/ or chemical properties and/ or their biological activity.

Liming Material – An inorganic soil conditioner containing one or both of the elements calcium and magnesium, generally in the form of an oxide, hydroxide, or carbonate, principally intended to maintain or raise the pH of soil.

Straight Fertilizer: A qualification generally given to a nitrogenous, phosphatic, or potassic fertilizer having a declarable content of only one of the primary plant nutrients, i.e. nitrogen, phosphorus, or potassium.

Compound Fertilizer: A fertilizer that has a declarable content of at least two of the plant nutrients nitrogen, phosphorus and potassium, obtained chemically or by blending or both.

Granular Fertilizer:– Solid material that is formed into particles of a predetermined mean size.

Coated Fertilizer – Granular fertilizer that is covered with a thin of a different material in order to improve the behavior and/ or modify the characteristics of the fertilizer.

Other related terms are:

Coated Slow- Release Fertilizer (AAPFCO)- A product containing sources of water- soluble nutrients, release of which in the soil is controlled by a coating applied to the fertilizer.

Polymer- Coated Fertilizer (AAPFCO)-A coated slow-release fertilizer consisting of fertilizer particles coated with a polymer (plastic) resin. It is a source of slowly available plant nutrients.

Controlled- Release Fertilizers- Fertilizers in which one or more of the nutrients have limited solubility in the soil solution, so that they become available to thea growing plant over a controlled period.

Nitrogen Stabilizer (AAPFCO) - A substance added to a fertilizer to extend the time that the nitrogen component of the fertilizer remains in the soil in the ammonia cal form.

Liquid Fertilizer – A term used for fertilizers in suspension or solution and for liquefied ammonia (ISO).

Solution Fertilizers (ISO) – Liquid fertilizer free of solid particles.

Suspension Fertilizer (ISO) – A two-phase fertilizer in which solid particles are maintained in suspension in the aqueous phase.

Suspension Fertilizer (AAPFCO) – A fluid containing dissolved and UN dissolved plant nutrients. The suspension of the undissolved plant nutrients may be inherent with the materials or produced with the aid of a suspending agent of nonfertilizer properties. Mechanical agitation may be necessary in some cases to facilitate uniform suspension of undissolved plant nutrients.

Suspension Fertilizer – A liquid (fluid) fertilizer containing solids held in suspension, for example, by the addition of a small amount of clay. The solids may be water-soluble in a saturated solution, or they may be insoluble, or both.

Slurry Fertilizer (AAPFCO)– A fluid mixture that contains dissolved and undissolved plant nutrient materials and requires continuous mechanical agitation to assure homogeneity.

Powder – A solid substance in the form of very fine particles. Powder is also referred to as "no granular fertilizer" and is sometimes defined as a fertilizer containing fine particles, usually with some upper limit such as 3 mm nut no lower limit.

Formula – A term used in some countries to express, by numbers, in the order N-P-K (nitrogen-phosphorus- potassium), the respective content of these nutrients in a compound fertilizer.

Bulk – Qualification given to a fertilizer or soil conditioner not packed in a container (ISO).

Guarantee (of Composition) – Quantitative and/ or qualitative characteristic with which a market product must comply for contractual or legal requirements.

Declarable – Content – That content of an element (or an oxide) which, according to national legislation, may be given on a label or document associated with a fertilizer or soil conditioner.

Fertilizer unit – The unit mass of a fertilizer nutrient (in the form of the element or an oxide) generally I kg.

Plant Food Ratio – The ratio of the numbers of fertilizer units in a given mass of fertilizer expressed in the order N – P – K.

Soil Chemistry

Soil chemistry is the study of the chemical characteristics of soil. Soil chemistry is affected by mineral composition, organic matter and environmental factors.

History

Until the late 1960s, soil chemistry focused primarily on chemical reactions in the soil that contribute to pedogenesis or that affect plant growth. Since then, concerns have grown about environmental pollution, organic and inorganic soil contamination and potential ecological health and environmental health risks. Consequently, the emphasis in soil chemistry has shifted from pedology and agricultural soil science to an emphasis on environmental soil science.

Environmental Soil Chemistry

A knowledge of environmental soil chemistry is paramount to predicting the fate of contaminants, as well as the processes by which they are initially released into the soil. Once a chemical is exposed to the soil environment myriad chemical reactions can occur that may increase or decrease contaminant toxicity. These reactions include adsorption/desorption, precipitation, polymerization, dissolution, complexation and oxidation/reduction. These reactions are often disregarded by scientists and engineers involved with environmental remediation. Understanding these processes enable us to better predict the fate and toxicity of contaminants and provide the knowledge to develop scientifically correct, and cost-effective remediation strategies.

Concepts

- Anion and cation exchange capacity
- Soil pH
- Mineral formation and transformation processes
- Clay mineralogy
- Sorption and precipitation reactions in soil
- Oxidation-reduction reactions
- Chemistry of problem soils

Soil pH

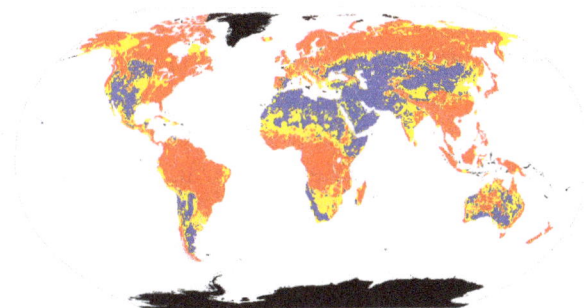

Global variation in soil pH. **Red** = acidic soil. **Yellow** = neutral soil. **Blue** = alkaline soil. **Black** = no data.

The soil pH is a measure of the acidity or alkalinity in soils. pH is defined as the negative logarithm (base 10) of the activity of hydronium ions (H^+ or, more precisely, $H_3O_{aq}^+$) in a solution. In soils, it is measured in a slurry of soil mixed with water (or a salt solution), and normally falls between

3 and 10, with 7 being neutral. A pH below 7 is acidic and above 7 is alkaline. Ultra-acidic soils (pH<3.5) and very strongly alkaline soils (pH>9) are rare.

Soil pH is considered a master variable in soils as it affects many chemical processes. It specifically affects plant nutrient availability by controlling the chemical forms of the different nutrients and influencing the chemical reactions they undergo. The optimum pH range for most plants is between 5.5 and 7.5; however, many plants have adapted to thrive at pH values outside this range.

Classification of Soil pH Ranges

The United States Department of Agriculture Natural Resources Conservation Service classifies soil pH ranges as follows:

Denomination	pH range
Ultra acidic	< 3.5
Extremely acidic	3.5–4.4
Very strongly acidic	4.5–5.0
Strongly acidic	5.1–5.5
Moderately acidic	5.6–6.0
Slightly acidic	6.1–6.5
Neutral	6.6–7.3
Slightly alkaline	7.4–7.8
Moderately alkaline	7.9–8.4
Strongly alkaline	8.5–9.0
Very strongly alkaline	> 9.0

Factors Affecting Soil pH

The pH of a natural soil depends in the mineral composition of the parent material of the soil, and the weathering reactions undergone by that parent material. In warm, humid environments, soil acidification occurs (soil pH decreases) over time as the products of weathering are leached by the flow of water through the soil. In dry climates, however, soil weathering and leaching are less intense and soil pH is often neutral or alkaline.

Sources of Acidity

Many processes contribute to soil acidification. These include:

- Rainfall: Acid soils are most often found in areas of high rainfall. Excess rainfall leaches base cation from the soil, increasing the percentage of Al^{3+} and H^+ relative to other cations.

Additionally, rainwater has a slightly acidic pH of 5.7 due to a reaction with CO_2 in the atmosphere that forms carbonic acid.

- Fertilizer use: Ammonium (NH_4^+) fertilizers react in the soil in a process called nitrification to form nitrate (NO_3^-,), and in the process release H^+ ions.

- Plant root activity: Plants take up nutrients in the form of ions (NO_3^-, NH_4^+, Ca^{2+}, $H_2PO_4^-$, etc.), and often, they take up more cations than anions. However plants must maintain a neutral charge in their roots. In order to compensate for the extra positive charge, they will release H^+ ions from the root. Some plants will also exude organic acids into the soil to acidify the zone around their roots to help solubilize metal nutrients that are insoluble at neutral pH, such as iron (Fe).

- Decomposition of organic matter by microorganisms releases CO_2 which when mixed with soil water can form carbonic acid (H_2CO_3).

- Acid rain: The burning of fossil fuels releases oxides of sulfur and nitrogen into the atmosphere, these react with water in the atmosphere to form sulfuric and nitric acid in rain.

- Oxidative weathering: Oxidation of some primary minerals, especially sulphides and those containing Fe^{2+} generate acidity. This process is often accelerated by human activity:

  o Mine spoil: Severely acidic conditions can form in soils near some mine spoils due to the oxidation of pyrite.

  o Acid sulfate soils formed naturally in waterlogged coastal and estuarine environments can become highly acidic when drained or excavated.

Sources of Alkalinity

Increased total soil alkalinity can occur with:

- Weathering of silicate, aluminosilicate and carbonate minerals containing Na^+, Ca^{2+}, Mg^{2+} and K^+;

- Addition of silicate, aluminosilicate and carbonate minerals to in soils; this may happen by deposition of material eroded elsewhere by wind or water, or by mixing of the soil with less weathered material (such as the addition of limestone to acid soils);

- Addition of water containing dissolved bicarbonates (as occurs when irrigating with high-bicarbonate waters).

The accumulation of alkalinity in a soil (as Na, K, Ca and Mg bicarbonates and carbonates) occurs when there is insufficient water flowing through the soils to leach soluble salts. This may be due to arid conditions, or poor internal soil drainage; in these situations most of the water that enters the soil is transpired (taken up by plants) or evaporates, rather than flowing through the soil.

The soil pH is usually increased when total alkalinity increases, but the balance of the added cations also has a marked effect on the soil pH – for example, increasing the amount of sodium in an alkaline soil will tend to induce dissolution of calcium carbonate, which will increase the pH.

Calcareous soils may vary in pH from 7.0 to 9.5, depending on the degree to which Ca^{2+} or Na^+ dominate the soluble cations.

Effect of Soil pH on Plant Growth

Acid Soils

Plants grown in acid soils can experience a variety of stresses including aluminium (Al), hydrogen (H), and/or manganese (Mn) toxicity, as well as nutrient deficiencies of calcium (Ca) and magnesium (Mg).

Aluminium toxicity is the most widespread problem in acid soils. Aluminium is present in all soils, but dissolved Al^{3+} is toxic to plants; Al^{3+} is most soluble at low pH; above pH 5.0, there is little Al in soluble form in most soils. Aluminium is not a plant nutrient, and as such, is not actively taken up by the plants, but enters plant roots passively through osmosis. Aluminium inhibits root growth; lateral roots and root tips become thickened and roots lack fine branching; root tips may turn brown. In the root, the initial effect of Al^{3+} is the inhibition of the expansion of the cells of the rhizodermis, leading to their rupture; thereafter it is known to interfere with many physiological processes including the uptake and transport of calcium and other essential nutrients, cell division, cell wall formation, and enzyme activity.

Proton (H^+ ion) stress can also limit plant growth. The proton pump, H^+-ATPase, of the plasmalemma of root cells works to maintain the near-neutral pH of their cytoplasm. A high proton activity (pH within the range 3.0–4.0 for most plant species) in the external growth medium overcomes the capacity of the cell to maintain the cytoplasmic pH and growth shuts down.

In soils with a high content of manganese-containing minerals, Mn toxicity can become a problem at pH 5.6 and lower. Manganese, like aluminium, becomes increasingly soluble as pH drops, and Mn toxicity symptoms can be seen at pH levels below 5.6. Manganese is an essential plant nutrient, so plants transport Mn into leaves. Classic symptoms of Mn toxicity are crinkling or cupping of leaves.

Nutrient Availability in Relation to Soil pH

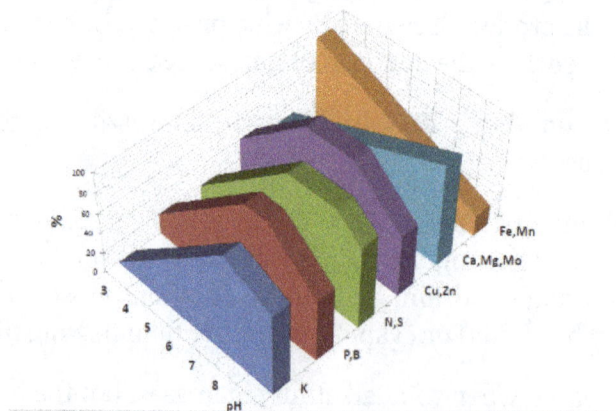

Nutrient availability in relation to soil pH

Soil pH affects the availability of some plant nutrients:

Aluminium toxicity reduces the availability of all nutrients by limiting root growth; this is largely limited to soil pH<5.0. Because roots are damaged, it becomes more difficult for plants to take up all nutrients, and deficiencies of the macronutrients (nitrogen, phosphorus, potassium, calcium and magnesium) are frequently encountered in very strongly acidic to ultra-acidic soils (pH<5.0).

Molybdenum availability is increased at higher pH; this is because the molybdate ion is more strongly sorbed by clay particles at lower pH.

Zinc, iron, copper and manganese show decreased availability at higher pH (increased sorbtion at higher pH).

The effect of pH on phosphorus availability varies considerably, depending on soil conditions and the crop in question. The prevailing view in the 1940s and 1950s was that P availability was maximized near neutrality (soil pH 6.5–7.5), and decreased at higher and lower pH. Interactions of phosphorus with pH in the moderately to slightly acidic range (pH 5.5–6.5) are, however, far more complex than this. Laboratory tests, glasshouse trials and field trials have indicated that increases in pH within this range may increase, decrease, or have no effect on P availability to plants.

Water Availability in Relation to Soil pH

Strongly alkaline soils are sodic and dispersive, with slow infiltration, low hydraulic conductivity and poor available water capacity. Plant growth is severely restricted because aeration is poor when the soil is wet; in dry conditions, plant-available water is rapidly depleted and the soils become hard and cloddy (high soil strength).

Many strongly acidic soils, on the other hand, have strong aggregation, good internal drainage, and good water-holding characteristics. However, for many plant species, aluminium toxicity severely limits root growth, and moisture stress can occur even when the soil is relatively moist.

Determining pH

Methods of determining pH include:

- Observation of soil profile: Certain profile characteristics can be indicators of either acid, saline, or sodic conditions. Strongly acidic soils often have poor incorporation of the organic surface layer with the underlying mineral layer. The mineral horizons are distinctively layered in many cases, with a pale eluvial (E) horizon beneath the organic surface; this E is underlain by a darker B horizon in a classic podzol horizon sequence. Presence of a caliche layer indicates the presence of calcium carbonates, which are present in alkaline conditions. Also, columnar structure can be an indicator of sodic condition.

- Observation of predominant flora. Calcifuge plants (those that prefer an acidic soil) include *Erica, Rhododendron* and nearly all other Ericaceae species, many birch (*Betula*), foxglove (*Digitalis*), gorse (*Ulex* spp.), and Scots Pine (*Pinus sylvestris*). Calcicole (lime loving) plants include ash trees (*Fraxinus* spp.), honeysuckle (*Lonicera*), *Buddleja*, dogwoods (*Cornus* spp.), lilac (*Syringa*) and *Clematis* species.

- Use of an inexpensive pH testing kit, where in a small sample of soil is mixed with indicator solution which changes colour according to the acidity/alkalinity.

- Use of litmus paper. A small sample of soil is mixed with distilled water, into which a strip of litmus paper is inserted. If the soil is acidic the paper turns red, if alkaline, blue.

- Use of a commercially available electronic pH meter, in which a glass or solid-state electrode is inserted into moistened soil and measures the hydrogen ion activity.

Plant pH Preferences

In general terms, different plant species are adapted to soils of different pH ranges. For many species, the suitable soil pH range is fairly well known. Online databases of plant characteristics, such *USDA PLANTS* and *Plants for a Future* can be used to look up the suitable soil pH range of a wide range of plants. Documents like *Ellenberg's indicator values for British plants* can also be consulted.

However, a plant may be intolerant of a particular pH in some soils as a result of a particular mechanism, and that mechanism may not apply in other soils. For example, a soil low in molybdenum may not be suitable for soybean plants at pH 5.5, but soils with sufficient molybdenum allow optimal growth at that pH. Similarly, some calcifuges (plants intolerant of high-pH soils) can tolerate calcareous soils if sufficient phosphorus is supplied. Another confounding factor is that different varieties of the same species often have different suitable soil pH ranges. Plant breeders can use this to breed varieties that can tolerate conditions that are otherwise considered unsuitable for that species - examples are projects to breed aluminium-tolerant and manganese-tolerant varieties of cereal crops for food production in strongly acidic soils.

The table below gives suitable soil pH ranges for some widely cultivated plants as found in the *USDA PLANTS Database*. Some species (like *Pinus radiata* and *Opuntia ficus-indica*) tolerate only a narrow range in soil pH, whereas others (such as *Vetiveria zizanioides*) tolerate a very wide pH range.

Scientific name	Common name	pH (minimum)	pH (maximum)
Vetiveria zizanioides	vetivergrass	3.0	8.0
Pinus rigida	pitch pine	3.5	5.1
Rubus chamaemorus	cloudberry	4.0	5.2
Ananas comosus	pineapple	4.0	6.0
Coffea arabica	Arabian coffee	4.0	7.5
Rhododendron arborescens	smooth azalea	4.2	5.7
Pinus radiata	Monterey pine	4.5	5.2
Carya illinoinensis	pecan	4.5	7.5
Tamarindus indica	tamarind	4.5	8.0
Vaccinium corymbosum	highbush blueberry	4.7	7.5

Scientific name	Common name	pH (minimum)	pH (maximum)
Manihot esculenta	cassava	5.0	5.5
Morus alba	white mulberry	5.0	7.0
Malus	apple	5.0	7.5
Pinus sylvestris	Scots pine	5.0	7.5
Carica papaya	papaya	5.0	8.0
Cajanus cajan	pigeonpea	5.0	8.3
Pyrus communis	common pear	5.2	6.7
Solanum lycopersicum	garden tomato	5.5	7.0
Psidium guajava	guava	5.5	7.0
Nerium oleander	oleander	5.5	7.8
Punica granatum	pomegranate	6.0	6.9
Viola sororia	common blue violet	6.0	7.8
Caragana arborescens	Siberian peashrub	6.0	9.0
Cotoneaster integerrimus	cotoneaster	6.8	8.7
Opuntia ficus-indica	Barbary fig (pricklypear)	7.0	8.5

Changing Soil pH

Increasing pH of Acidic Soil

Finely ground agricultural lime is often applied to acid soils to increase soil pH (liming). The amount of lime needed to change pH is determined by the mesh size of the lime (how finely it is ground) and the buffering capacity of the soil. A high mesh size (60 mesh = 0.25 mm; 100 mesh = 0.149 mm) indicates a finely ground lime that will react quickly with soil acidity. The buffering capacity of a soil depends on the clay content of the soil, the type of clay, and the amount of organic matter present, and may be related to the soil cation exchange capacity. Soils with high clay content will have a higher buffering capacity than soils with little clay, and soils with high organic matter will have a higher buffering capacity than those with low organic matter. Soils with higher buffering capacity require a greater amount of lime to achieve an equivalent change in pH.

Amendments other than agricultural lime that can be used to increase the pH of soil include wood ash, industrial calcium oxide (burnt lime), magnesium oxide, basic slag (calcium silicate), and oyster shells. These products increase the pH of soils through various acid-base reactions. Calcium silicate neutralizes active acidity in the soil by reacting with H^+ ions to form monosilicic acid (H_4SiO_4), a neutral solute.

Decreasing pH of Alkaline Soil

The pH of an alkaline soil can be reduced by adding acidifying agents or organic materials such as the ones listed below. Acidifying fertilizers, such as those containing ammonium, can help to reduce the pH of a soil. However, in high-pH soils with a high calcium carbonate content (more than 2%), it can be very costly and/or ineffective to attempt to reduce the pH with acids. In such cases, it is often more efficient to add phosphorus, iron, manganese, copper and/or zinc instead, because deficiencies of these nutrients are the most common reasons for poor plant growth in calcareous soils.

- Amend the soil with organic matter. On average, soils with higher organic matter contents have lower pH. Peat or sphagnum peat moss are highly acidic and will lower soil pH more than other organic amendments.

- Add elemental sulfur (90-99% S) annually at a rate of 300-500 kg/ha (6 to 10 pounds per 1000 square feet). Elemental sulfur slowly oxidizes in soil to form sulfuric acid. Test the soil occasionally and stop adding sulfur when pH has reached desirable levels.

- Use acidifying fertilizers such as ammonium sulfate and other products with label designations indicating an acidic reaction in the soil. With repeated use these materials often reduce soil pH.

- Plant on raised beds in a sandy medium amended with peat moss or another source of acidic organic matter. An alternative is to plant in boxes or half-barrels heavily amended with acidic forms of organic matter.

Soil Acidification

Soil acidification is the buildup of hydrogen cations, also called protons, reducing the soil pH. This happens when a proton donor gets added to the soil. The donor can be an acid, such as nitric acid and sulfuric acid (these acids are common components of acid rain). It can also be a compound such as aluminium sulfate, which reacts in the soil to release protons. Many nitrogen compounds, which are added as fertilizer, also acidify soil over the long term because they produce nitrous and nitric acid when oxidized in the process of nitrification.

Acidification also occurs when base cations such as calcium, magnesium, potassium and sodium are leached from the soil. This leaching increases with increasing precipitation. Acid rain accelerates the leaching of bases. Plants take bases from the soil as they grow, donating a proton in exchange for each base cation. Where plant material is removed, as when a forest is logged or crops are harvested, the bases they have taken up are permanently lost from the soil.

Plant Leaves Left on Soil

Many plants produce organic acids. Where plant litter accumulates on or is incorporated to the soil, these acids (including acetic acid, humic acid, oxalic acid, and tannic acid) are liberated. This is especially acute in soils under coniferous trees such as pine, spruce and fir, which return fewer base cations to the soil than do most deciduous trees.

Rocks in the Soil

Certain parent materials also contribute to soil acidification. Granites and their allied igneous rocks are called "acidic" because they have a lot of free quartz, which produces silicic acid on weathering. Also, they have relatively low amounts of calcium and magnesium. Some sedimentary rocks such as shale and coal are rich in sulfides, which, when hydrated and oxidized, produce sulfuric acid which is much stronger than silicic acid. Many coal spoils are too acidic to support vigorous plant growth, and coal gives off strong precursors to acid rain when it is burned. Marine clays are also sulfide-rich in many cases, and such clays become very acidic if they are drained to an oxidizing state.

Pollution

Acidification may also occur from nitrogen emissions into the air, as the nitrogen may end up deposited into the soil.

Acidifying Compounds

- Aluminium sulfate
- Ammonia
- Ammonium nitrate
- Ammonium phosphate
- Ammonium sulfate
- Ferrous sulfate
- Monopotassium phosphate
- Phosphoric acid
- Urea
- Alum

Classifications of Soil Nutrients

Only a few forms of nutrients are taken up by plants. Concentrations of these forms of nutrients are normally low in soils, but soils have the ability to store nutrients in many organic and inorganic molecules. They are transformed to the soluble plant available forms at rates defined by the environmental conditions for the biotic and abiotic reactions, amounts of the different non available forms, their rates of conversion, and removal of various nutrient forms from soil solution. Thus, all of the plant nutrients have ionic forms that are plant available, and soils contain an assortment of less available forms that became available to plants through chemical and biological reactions.

The nutrients supplied from soil have been classified in different ways: by their functions in plants, by whether they generally need to be added to soils to optimize growth, or by their mobility in the plant. For agricultural production, nutrients are classified by the quantities of the nutrients needed by plants and whether they are normally present in sufficient quantities in soils to optimize the growth of a crop.

Macronutrients

The elements nitrogen (N), phosphorus (P), and potassium (K) are the nutrients that most often limit crop growth, and they are called "macronutrients." Since N, P, and K are the nutrients most widely deficient in soils, they are also called the "major," "primary," or the "fertilizer" nutrients.

Calcium, Mg, and S are called the "secondary" nutrients of the macronutrients because they are not as widely deficient as N, P, and K.

Macronutrients	Chemical Symbol	Principal forms taken up by plants
Primary (Major)		
Nitrogen	N	NH_4+ and NO_3
Phosphorus	P	$HPO_4=$ and H_2PO_4-
Potassium	K^b	K^+
Secondary (Minor)		
Calcium	Ca	Ca^{++}
Magnesium	Mg	Mg^{++}
Sulfur	S	HSO_4- and $SO_4=$
Macronutrients (Trace Elements)		
Boron	B	H_3BO_3, H_2BO_3-
Copper	C	$Cu^{++}, Cu(OH)^+$
Chlorine	Cl	Cl^-
Iron	Fe	$Fe^{++}, Fe(OH)_2^+, Fe(OH)^{++}, Fe^{+++}$
Manganese	Mn	Mn^{+2}
Molybdenum	Mo	$M_0O_4=, HM_0O_4-$
Zinc	Zn	Zn^{++}
Cobalt	Co	Co^{++}

Micronutrients

Other elements that are needed by plants in much lower amounts than the macronutrients are called micronutrients or trace elements. They are as essential to plant growth as the macronutrients because they perform very essential and specific roles, particularly in molecules involved with energy transfer process, hormones, and enzymes.

Fundamentals of N, P and K

Soil factors that affect the plant availability of the primary or fertilizer nutrients (N, P and K) are quite complex, and different from one another. The available forms, mechanisms of becoming plant available, forms of stored nutrients in soils, and timing of plant uptake are different not only between the nutrients but also for individual fertilizer sources of each nutrient. A basic understanding of the reaction of nutrients in soils and the factors that affect the response of crops to fertilizers is essential for people in the fertilizer industry so that they can make informed production and marketing decisions based on what can increase the profit from using fertilizers.

Soil Nitrogen

Nitrogen is a component of amino acids, which make up proteins, chlorophyll (the molecule that captures the sun's energy), enzymes, and the genetic material, nucleic acids. Therefore, this nutrient is needed in large amounts by all plants. Plants that do not have adequate N are yellowish, have yellowing and browning older leaves, are stunted and have poor root systems. Without additions of N, the nutrients are practically uniformly deficient for all grasses and cultivated crops. The only exceptions are leguminous crops that have symbiotic bacterial colonies growing within the plant, which have the ability to meet N needs through fixation of atmospheric N_2 Excess N can cause excessive vegetative growth, delayed maturity, and stalk breakage in small grains. The other macronutrients do not have direct detrimental effects when applied to excess.

The availability of N to plants is largely controlled by soil microbial processes. The N cycle in soils is complex, and under certain conditions large amounts of plant available N can be lost from the soil in drainage water or to the atmosphere. In this way, N is different from the other nutrients, which are not as readily lost from soils. The rates of conversion between forms of N and the direction (immobilization and mineralization) depend principally on the growth conditions of the microorganisms, i.e., temperature, moisture, oxygen availability and composition of organic substrates etc, rather than simply abiotic reactions.

Nitrogen uptake depends on the relatively small amount of available N forms in soil solution the soil's ability to replenish the available forms, and the growth conditions for the plant. Of course, any condition that inhibits plant growth such as other nutrient deficiencies, poor rooting conditions, poor weather, etc., will reduces N uptake.

Nitrate (NO_3^-) is the main form of N available to upland crops, whereas ammonium (NH_4^+) is the main form taken up by plants growing under flooded conditions, notably rice. Nitrate moves through soils with the soil water, since soils have little anion-absorbing capacity. Therefore, NO_3^- can move with water to plant roots for uptake. Nitrate is also eligible for leaching into groundwater. To reduce NO_3^- pollution, it is therefore important that the amount of NO_3^- stored in the soil be minimized during periods in which water percolates through the soil. Fertilizer N applications can be made more efficient by applying the N before it is needed by the crop.

Nitrate-N is also subject to loss from the soil through microbial processes when soils become highly anaerobic due to water saturation, which inhibits oxygen movement. When soils become anaerobic, certain microbes use NO_3^- as a terminal election acceptor, and the NO_3^- is converted to N_2 and N_2o, forms of N that cannot be used by plants. This loss of plant available N, called denitrification, therefore precludes NO_3^- use in flooded rice culture. When soils are flooded to grow rice, the NO_3^- accumulated from organic matter transformations during the period of dry season fallow is lost through denitrification. Upland soils, which become very wet for a week or more because of poor internal drainage or weather conditions, can lose NO_3^- through denitrification, particularly if there is readily degradable organic material in the soil. The organic material causes O_2 to be consumed more rapidly and provides growth substances for the denitrification organisms.

While the denitrification process is lamented by agronomists as a loss of plant available N from the soil system, it is the only way that N is recycled back to the atmosphere as N_2. Without this process, lower soil layers and ground waters would become large reservoirs of NO_3^- and the oxygen-enriched atmosphere would support continual conflagrations and thus, make life difficult. Therefore, the N cycle is just as important as the carbon and the hydrologic cycles to the life support system of the earth.

Ammonium N is a plant available from in addition to NO_3^-, but in normal upland soils it is converted fairly rapidly to NO_3^- by a series of microbial reactions collectively called nitrification. Most NH_4^+ fertilizers are converted to NO_3^- within a few weeks, but this period can vary with soil temperature, moisture, O_2 availability, soil pH, and the manner in which NH_4^+ or NH_3 was applied. Soils of less than 5.0 nitrify more slowly than more neutral soils. Concentrated zones of NH_4^+ can also reduce the rate of its conversion to NO_3^- As a cation, NH_4+ can be absorbed onto cation exchange sites and is therefore normally not subject to leaching as is NO_3^-. In addition, NH_4^+ can move into the interior of clay lattice structures and be protected from nitrification or plant uptake; this is called ammonium fixation.

N in organic forms is the major storehouse of N in soils and since the biological conversions occur fairly rapidly, plants rely mainly on N derived from organic material to meet their N needs. When NH_4^+ and NO_3^- are added to soils, the soil microorganisms make the major decisions as to what happens to the N, that is, whether the N is put into the organic N pool by their assimilation of the N or is converted to other forms of N either usable or unusable by plants. when a carbonaceous material, for instance straw, is added to soils, NO_3^- or NH_4^+ (termed mineral N) will be used by the microbes to utilize the carbon energy source, and there will be bet immobilization or incorporation of the plant available N into unavailable organic forms. This organic N can then be converted back to NH_4^+ and then to NH_3^- after only a few weeks and becomes available to the crop. This organic pool of N is therefore the most important aspect of N nutrition, and 70% - 80% of the N taken up by crops is normally from this pool rather than directly from mineral fertilizer additions. Synthesized fertilizer N sources not only increase the available mineral N status of soil but also temporarily increase the amount of N in soil organic matter as well.

Soil Phosphorus

Phosphorus compound are important in energy transfer and storage reactions since electrons are moved via adenosine di- and triphosphate in most biochemical reactions. Phosphorus is also a constituent of nucleic acids, which make up genetic encoded strands of DNA and RNA used for protein synthesis. Flowering, fruiting, seed formation, and early plant development therefore rely heavily on P supplies, as do root development and several other processes. Even though the P content of plant tissues is about 0.1% - 0.4%, which is about one-half of the amount of N or K, P is a critical component of particular basic to life. Phosphorus is fairly mobile in plants and is therefore moved to the youngest tissues of plants are P deficient. Deficiencies of P are not as readily observed as those of N, or even K, through the effects on growth and yield are often dramatic. Some plant species exhibit a purple or red coloration when they are deficient in P all will show overall stunting.

The orthophosphate form H_2PO_4 predominates as the form of P taken up by plants although there are also a few organic compounds that have been shown to be taken up directly plants. The solubility of orthophosphate compounds in the soil solution may vary from 0.05 to 3 ppmw. Unlike N, which is supplied as NO_3^- by mass flow of water to plant roots, only a fraction of the P needed by plants is supplied by mass flow with soil water. Diffusion is a much more important mechanism by which plant roots are supplied with P, and the amounts of P in solution and its replenishment are determined by the dissolution of relatively insoluble P compounds or minerals. However, plant roots are not totally passive sinks for P; the exudation of organic materials provides substrates for microbial populations, which produce various acids that help to dissolve P compounds in soils. Zones around plant roots, called the rhizosphere, support colonies of microorganisms that are essential to the growth of many species of plants.

Essential all native soil P originates from apatitic minerals. As apatites weather, new minerals are formed, which are relatively soluble, and make up the labile P pool, which replenishes P in soil solution. Many less soluble forms are also formed that comprise a nonlabile pool, which is a large but fairly inert storehouse for P.

The P moves between the various pools of varying solubilities in response to changing phosphate and cation concentrations. The distribution of these compounds depends principally on the Ca, Fe, Mn and Al concentrations and the soil pH. Organic forms of P can also be of importance I soils. Young soils, which contain weather apatites, are better endowed with available P than older, more weathered and acidified soils.

Phosphorus Fixation

When water-soluble sources of P are added to soils, the orthophosphate forms are converted to many different compounds, depending on the above elements and soil pH. In soils of moderate pH, Ca phosphate dominates, whereas Fe and Al compounds become more important in more acidic soils. These compounds then revert to even more stable or soluble compounds with time. Various sorption reactions onto metal oxides; hydroxides, and carbonate mineral surfaces also occur; these reactions are either reversible or relatively irreversible and affect phosphate solubilities and supply. These relatively irreversible reactions of phosphate are collectively called P fixation and are very dependent on the chemical constituents of soils. Soils that are of near neutral reaction and sandy texture have little fixation; next would come other neutral soils with lower levels of soluble Fe and Al. soils with the highest fixation are acidic, clayey soils high in Fe and Al. younger soils tend to have more available-P and lower fixation capacity although some young volcanic soils have colloidal amorphous materials that fix large amounts of P. the response of crops to added water soluble phosphate fertilizers is very dependent on P fixation capacity since the amount of P which is actually available depends on the P fixation capacity .

Figure attempts to illustrate several points (1) crop response to P fertilizer is very dependent on the fixation capacity of the soil that is fertilized; (2) response to P can change dramatically with P fertilization after the fixation capacity is satisfied; (3) after P fixation is satisfied and the available P level is raised to the no response zone, further P additions are only needed to replenish P removed by the crop; and (4) the efficiency of P applications will be different on the same soil, depending on the amount of soluble P fertilizer that have been previously applied.

A favorable aspect of the mechanisms by which P is held and released in soils is that there is little P lost from a soil except through harvested crop removal and soil erosion. The P lost through erosion is contained in clay particles and organic matter. There generally is little movement of P through soils with percolating water since soils contains Ca, Fe and Al in sufficient quantity to precipitate the forms of P. However, extremely P fertile soils can leach small amounts of ortho-phosphate. Organic-P forms are more mobile in soils than mineral forms. Soils that have had large amounts of animal manures added can have organic-P compounds leached into relatively shallow water tables. The P added in animal manures can move perhaps twice as far as similar amounts of inorganic P.

In fortunate contrast to N, the formation of P compounds of limited solibilities in soils allows P fertility to increase through P additions and P applied in one year can furnish P to succeeding crops. This is called a residual fertilizer effect. The P not used by the crop remains mainly in the nonlabile or plant-unavailable pool, but it still contributes to the P nutrition of crops for many years. Form soley an agronomic perspective, it is most beneficial to build P fertility rapidly by fairly large addition, and then apply the amounts needed to maintain the P status, which can be done every other year or every third year.

Soil Potassium

Potassium is considerably different than the other fertilizer nutrients, both in its function within plants and in the way it becomes available to plants. Unlike most nutrients, K in plants does not become part of structural components of plants nor elements of proteins, carbohydrates, etc., but it remains as an ion, mainly in cellular liquids. It functions to control turgidity of cells, helps in the transport mechanisms of starches and sugar, acts in protein synthesis, and activates enzymes. It encourages root growth, makes plant stalks healthy and strong, and is very important in reducing the effects of drought. Frequently, the most important benefits of good K nutrition are not in yield but also from the reduction of fungal disease and improved quality of harvested products. Since K is very mobile in plants, deficiency symptoms are seen in older leaves first, usually as yellowing or browning of the leaf margins, or in some crops as brown spots on the older leaves. Deficiency symptoms are not as readily observable as those of N, even though K is needed in almost the same amounts as N. overall slow growth, lodging(the plant falls over from stalk weakness), and poor grain or fruit quality are often indicators in roots and grain development and many other aspects of plant growth and development.

As a cation, K enters soil solution from cation exchange sites of clays and organic materials. The exchangeable K originates from the weathering of the minerals contained in the rocks from which soils are formed. Potassium feldspars are very natural reserves of available K in young soils; however, they are rapidly weathered and thus are essentially absent from moist tropical soils. Various micas, orthoclase and microcline are primary minerals degrade to illite, clay chlorite, and vermiculite with release of K^+ contained as layer in the clay lattices. Vermiculite and its degradation product montmorillonite (smectite) are 2:1 layer clays, meaning that two tetrahedral silica sheets around one octahedral alumina sheet compromise the basic structure. These clays soil moisture, an important characteristic because they can affect water movement and storage, K release and absorption and soil structure.

Young soils with appreciable clay content commonly contain large amounts of K and some of these soils can be cropped for years without developments of substantial K deficiencies. Weathered, sandy and organic soils are primarily in need of K additions. Sandy and weathered soils, particularly under warm, wet climatic conditions under which organic matter degrades rapidly, have low CEC, which means they can retain only small quantities of K. they also lack clays that can fix K within the clay lattices. Potassium added to such soils can be leached as can other base cations, particularly when other nutrients such as N and S are added, which add concomitant anions to maintain electro neutrality in the leachate.

The leaching of K^+ tends to be more dependent upon the amount of percolate passing through the soil than on soil texture. Sandy soils, however, can infiltrate more water than heavier soils, so less runoff occurs than with a heavier soil on a similar slope and rainfall intensity. Leaching losses of K^+ can vary from essentially nil to somewhat more than 100kg K ha-1. Losses from sandy irrigated soils can be particularly high. Soil liming generally decrease the amount of K leached from soils; in one case it reduced losses to about one half , and this is an important aspect of limiting in maintaining soil fertility.

Losses of K through erosion of clays from the soil surface can be very substantial. Few measurements of runoff loss have apparently been done, but these losses are certainly important.

Since K can be lost from soils and fixed in clay minerals, yearly additions of K are more efficient than larger, infrequent applications on some soils although few crops respond to split applications of K in the same season as they do to N.

Plants need large amounts of K, which is removed in harvested plant parts. Plant residues left in the field can return some K back to the soil. Animal manures are important sources of K because they often contain as much K as N. applied K salts, as KCl or K_2SO_4, give the same K response, but in some cases the crops may respond to either the Cl or S carriers.

Soil Sulfur

Sulfur is needed by plants in amount similar to those of P. It is an essential constituent of three amino acids, in building blocks of proteins and enzymes. It cycles in the soil-plant-atmospheric system similarly to N, with reduced sulfide forms (H_2S, pyrites), oxidized sulfate dioxide and organic sulfur compounds exist in the atmosphere. As with nitrogen, organic forms of S comprise the major storehouse for soil S. A notate difference between N and S is that gypsum ($CaSo_4 2H_2O$) a relatively insoluble mineral, can form in soil to maintain S in more arid soils. Insoluble sulfates also form in wetter soils and accumulate S in subsoils. Obviously the atmosphere does not contain S as it does N_2.

Sulfur deficiencies in plants look similar and are often confused with those of N. both deficiencies are manifest as yellowing of the plant; yet, there are clear distinctions. Deficiencies of S are seen in the younger leaves, whereas those of N are seen mainly in the older leaves. Sulfur deficiencies result in small, spindly plants and severely reduced nodulation for BNF in legumes. Seed and fruit maturation is also delayed.

Sulfur deficiencies can readily develop, particularly in humid areas and areas that do not receive S from the atmosphere or from S containing fertilizers. Atmospheric sources of S are organic S compounds emitted from marine plankton and inorganic S from oil and coal combustion. Because enforcement of environmental laws and better technologies have reduced the amount of S that soils receive through acid rain, this source has decreased. In addition, since higher analysis fertilizer have largely replaced S containing single superphosphate and ammonium sulfate, incidental contributions of S to soils are reduced. Sulfur deficiencies are therefore expected to increase and sulfate containing fertilizers will probably be needed, particularly in moist tropical areas and especially for legume and vegetable production.

Organic Fertilizer

A cement reservoir containing cow manure mixed with water. This is common in rural Hainan Province, China. Note the bucket on a stick that the farmer uses to apply the mixture.

Organic fertilizers are fertilizers derived from animal matter, animal excreta (manure), human excreta, and vegetable matter. (e.g. compost and crop residues). Naturally occurring organic fertilizers include animal wastes from meat processing, peat, manure, slurry, and guano.

In contrast, the majority of fertilizers used in commercial farming are extracted from minerals (e.g., phosphate rock) or produced industrially (e.g., ammonia). Organic agriculture, a system of farming, allows for certain fertilizers and amendments and disallows others; that is also distinct from this topic.

Examples and Sources

The main organic fertilizers are, peat, animal wastes (often from slaughter houses), plant wastes from agriculture, and treated sewage sludge.

Mineral

Peat is the most widely used organic amendment.

By some definitions, minerals are distinctly separate from organic materials. However, certain organic fertilizers and amendments are mined, specifically guano and peat, and other mined minerals are fossil products of animal activity, such as greensand (anaerobic marine deposits), some limestones (fossil shell deposits) and some rock phosphates, (fossil guano). Peat, a precursor to coal, offers no nutritional value to the plants, but improves the soil by aeration and absorbing water; it is sometimes credited as being the most widely used organic fertilizer and by volume is the top organic amendment.

Animal Sources

These materials include the products of the slaughter of animals. Bloodmeal, bone meal, hides, hoofs, and horns are typical precursors. fish meal, and feather meal are other sources.

Chicken litter, which consists of chicken manure mixed with sawdust, is an organic fertilizer that has been shown to better condition soil for harvest than synthesized fertilizer. Researchers at the Agricultural Research Service (ARS) studied the effects of using chicken litter, an organic fertilizer, versus synthetic fertilizers on cotton fields, and found that fields fertilized with chicken litter had a 12% increase in cotton yields over fields fertilized with synthetic fertilizer. In addition to higher yields, researchers valued commercially sold chicken litter at a $17/ton premium (to a total valuation of $78/ton) over the traditional valuations of $61/ton due to value added as a soil conditioner.

Plant

Processed organic fertilizers include compost, humic acid, amino acids, and seaweed extracts. Other examples are natural enzyme-digested proteins. Decomposing crop residue (green manure) from prior years is another source of fertility.

Other ARS studies have found that algae used to capture nitrogen and phosphorus runoff from agricultural fields can not only prevent water contamination of these nutrients, but also can be used as an organic fertilizer. ARS scientists originally developed the "algal turf scrubber" to reduce nutrient runoff and increase quality of water flowing into streams, rivers, and lakes. They found that this nutrient-rich algae, once dried, can be applied to cucumber and corn seedlings and result in growth comparable to that seen using synthetic fertilizers.

Treated Sewage Sludge

Although night soil (from human excreta) was a traditional organic fertilizer, the main source of this type is nowadays treated sewage sludge, also known as biosolids.

Decomposing animal manure, an organic fertilizer source

Biosolids as soil amendment is only available to less than 1% of US agricultural land. Industrial pollutants in sewage sludge prevents recycling it as fertilizer. The USDA prohibits use of sewage sludge in organic agricultural operations in the U.S. due to industrial pollution, pharmaceuticals, hormones, heavy metals, and other factors. The USDA now requires 3rd-party certification of high-nitrogen liquid organic fertilizers sold in the U.S.

Sewage sludge use in organic agricultural operations in the U.S. has been extremely limited and rare due to USDA prohibition of the practice (due to toxic metal accumulation, among other factors).

Urine

Animal sourced urea and urea-formaldehyde from urine are suitable for organic agriculture; however, synthetically produced urea is not. The common thread that can be seen through these examples is that *organic* agriculture attempts to define itself through minimal processing (e.g., via chemical energy such as petroleum), as well as being naturally occurring or via natural biological processes such as composting.

Biosolids

Pumpkin seedlings planted out on windrows of composted biosolids

Biosolids is a term coined in the United States that is typically used to describe several forms of treated sewage sludge that is intended for agricultural use as a soil conditioner. Although sewage sludge has long been used in agriculture, concerns about offensive odors and disease risks from pathogens and toxic chemicals may reduce public acceptance of the practice. Modern use of the term *biosolids* may be subject to government regulations, although informal use describes a broader range of semi-solid organic products separated from sewage.

Description of *biosolids* in conformance with local regulations may reduce confusion; but some use an expanded definition including any solids, slime solids or liquid slurry residue generated during the treatment of domestic sewage including scum and solids removed during primary, secondary or advanced treatment processes. Use of alternative terms like *solids* or *wastewater solids* may be preferable for non-conforming biosolids.

Terminology

Biosolids may be defined as organic wastewater solids that can be reused after suitable sewage sludge treatment processes leading to sludge stabilization such as anaerobic digestion and composting.

Alternatively, the biosolids definition may be restricted by local regulations to wastewater solids only after those solids have completed a specified treatment sequence and/or have concentrations of pathogens and toxic chemicals below specified levels.

The United States Environmental Protection Agency (USEPA) defines the two terms - sewage sludge and biosolids - in the Code of Federal Regulations (CFR), Title 40, Part 503 as follows: *sewage sludge* refers to the solids separated during the treatment of municipal wastewater (including domestic septage), while *biosolids* refers to treated sewage sludge that meets the USEPA pollutant and pathogen requirements for land application and surface disposal. A similar definition has been used internationally.

Characteristics

Quantities

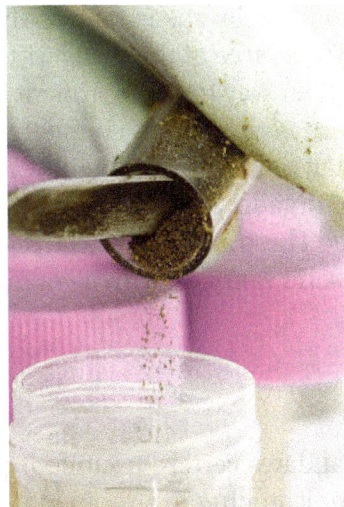

Testing for human pathogens in cereal crops after the application of biosolids. Biosolids are applied as fertilizer in the Central Wheatbelt of Australia as a recycling program by the Water Corporation.

Approximately 7,100,000 dry tons of biosolids were generated in 2004 at approximately 16,500 municipal wastewater treatment facilities in the United States.

In the United States, as of 2013 about 55% of sewage solids are turned into fertilizer, despite demand from farmers who wish to buy more. Challenges to increased levels of recycling include capital needed to build digesters, the complexity of complying with health regulations, and avoiding neighbors who object to unpleasant smells. There are also new forms of contaminants in urban sewage systems which make the process of producing high quality biosolids more complex. These have led some municipalities to ban biosolids on farms and even in forests.

Nutrients

Encouraging agricultural use of biosolids is intended to prevent filling landfills with nutrient-rich organic materials from the treatment of domestic sewage that might be recycled and applied as fertilizer to improve and maintain productive soils and stimulate plant growth. Biosolids may contain macronutrients nitrogen, phosphorus, potassium and sulphur with micronutrients copper, zinc, calcium, magnesium, iron, boron, molybdenum and manganese.

Industrial and Man-made Contaminants

The United States Environmental Protection Agency (USEPA) and others have shown that biosolids can contain measurable levels of synthetic organic compounds, radionuclides and heavy metals. USEPA has set numeric limits for arsenic, cadmium, copper, lead, mercury, molybdenum, nickel, selenium, and zinc but has not regulated dioxin levels.

Contaminants from pharmaceuticals and personal care products and some steroids and hormones may also be present in biosolids. Substantial levels of persistent, bioaccumulative and toxic (PBT) polybrominated diphenyl ethers were detected in biosolids in 2001. This finding came despite the EPA's previous assertion that all PBT organic pollutants of concern had been banned from production in the 1970s and hence these could be ignored in risk assessment. In 2007 toxic PCBs were detected in the biosolid product Milorganite, donated to the City of Milwaukee and subsequently applied on city parkland. The cost to the Milwaukee Metropolitan Sewerage District and tax payers was estimated as $4.7 million. The source of the PCBs was later determined to be a shuttered die-casting facility. The PCBs made their way to the treatment plant via sewer lines years after the facility stopped operation. PCBs were banned from commerce in the US in the mid-1970s. The United States Geological Survey analyzed in 2014 nine different consumer products containing biosolids as a main ingredient for 87 organic chemicals found in cleaners, personal care products, pharmaceuticals, and other products. These analysis detected 55 of the 87 organic chemicals measured in at least one of the nine biosolid samples, with as many as 45 chemicals found in a single sample.

Pathogens

In the United States the USEPA mandates certain treatment processes designed to significantly decrease levels of certain so-called indicator organisms, in biosolids. These include, "...operational standards for fecal coliforms, *Salmonella* sp. bacteria, enteric viruses, and viable helminth ova."

However, the US-based Water Environment Research Foundation has shown that some pathogens do survive sewage sludge treatment.

USEPA has also classified other pathogens that can appear in biosolids such as various protozoa, bacteria, viruses, and prions as "pathogens of emerging concern".

EPA regulations allow only biosolids with no detectable pathogens to be widely applied; those with remaining pathogens are restricted in use. Unfortunately, prions isn't on the list of pathogens that it regulates. The risk assessments are outdated, since prions were not recognized by science until the 1990s. The EPA's infamous sludge rule was crafted several years earlier.

Classification Systems

United States

In the United States Code of Federal Regulations (CFR), Title 40, Part 503 governs the management of biosolids. Within that federal regulation biosolids are generally classified differently depending upon the quantity of pollutants they contain and the level of treatment they have been subjected to (the latter of which determines both the level of vector attraction reduction and the level of pathogen reduction). These factors also affect how they may be disseminated (bulk or bagged) and the level of monitoring oversight which, in turn determines where and in what quantity they may be applied.

History

As public concern arose about disposal in the United States of increasing volumes of solids being removed from sewage during sewage treatment mandated by the Clean Water Act, the Water Environment Federation (WEF) sought a new name to distinguish the clean, agriculturally viable product generated by modern wastewater treatment from earlier forms of sewage sludge widely remembered for causing offensive or dangerous conditions. Of three-hundred suggestions, *biosolids* was attributed to Dr. Bruce Logan of the University of Arizona, and recognized by WEF in 1991.

Microbiologist and EPA whistleblower, David. L. Lewis, has documented illness, death and livestock destruction traced to the use of biosolids. He also charges that the National Academy of Sciences, EPA, USDA, and other vested interests have expunged documentation and studies from reports in order to protect the EPA policy of using biosolids. Furthermore, Dr. Lewis charges that the emphasis on using biosolids in low-income urban and rural settings especially in the early years of "sludge magic" is an Environmental Injustice and human experimentation without informed consent.

Examples

- Milorganite is the trademark of a biosolids fertilizer produced by the Milwaukee Metropolitan Sewerage District. The recycled organic nitrogen fertilizer from the Jones Island Water Reclamation Facility in Milwaukee, Wisconsin, is sold throughout North America, reduces the need for manufactured nutrients.

- Loop is the trademark of a biosolids soil amendment produced by the King County Wastewater Treatment Division. Loop has been blended into GroCo, a commercially available compost product, since 1976. Several local farms and forests also use Loop directly.

- TAGRO is short for "Tacoma Grow" and is produced by the City of Tacoma, Washington since 1991.

- Dillo Dirt has been produced by the City of Austin, Texas since 1989.

Reuse of Excreta

Reuse of excreta (or re-use or use of excreta) refers to the safe, beneficial use of animal or human excreta, i.e. feces (or faeces in British English) and urine. Such beneficial use involves mainly the nutrient, organic matter and energy contained in excreta, rather than the water content which is the case for wastewater reuse. Reuse of excreta can involve using it as soil conditioner or fertilizer in agriculture, gardening, aquaculture or horticultural activities. Other uses include use as a fuel source, building material or for protein food production.

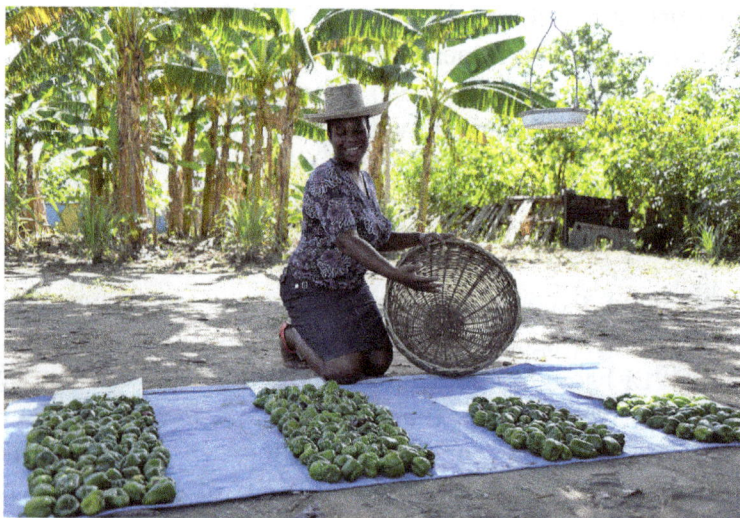

Harvest of peppers grown with excreta-based compost at an experimental garden of SOIL in Haiti

Excreta contains resources that can be recovered: plant-available nutrients nitrogen, phosphorus, potassium as well as micronutrients such as sulphur and organic matter. These resources which are contained in excreta or in domestic wastewater (sewage) have traditionally been used in agriculture in many countries. They are still being used in agriculture to this day, but the practice is often carried out in an unregulated and unsafe manner in developing countries. The WHO Guidelines from 2006 have set up a framework how this reuse can be done safely by following a "multiple barrier approach".

There are a number of "excreta-derived fertilizers" which vary in their properties and fertilizing characteristics: urine, dried feces, composted feces, fecal sludge (septage), sewage, sewage sludge and animal manure.

Reuse of excreta is the final step of the sanitation chain which starts with collection of excreta (by use of toilets) and continues with transport and treatment all the way to either disposal or reuse.

Background

Sanitation systems that are designed for safe, effective recovery of resources can play an important role in a community's overall resource management. Various technologies and practices, ranging in scale from a single rural household to a city, can be used to capture potentially valuable resources and make them available for safe, productive uses that support human well-being and broader sustainability.

The resources available in wastewater and excreta include water, plant nutrients, organic matter and energy content. Reuse of excreta focuses on the nutrient and organic matter content of excreta unlike reuse of wastewater which focuses on the water content.

The most common type of reuse of excreta is as fertilizer and soil conditioner in agriculture. This is also called a "closing the loop" approach for sanitation with agriculture. It is a central aspect of the ecological sanitation approach. An alternative term is also "use of excreta" rather than "reuse" as strictly speaking it is the *first use* of excreta, not the second time that it is used.

It can be efficient to combine wastewater and excreta with other organic waste such as manure, food and crop waste for the purposes of resource recovery.

Types

The most common types of excreta reuse, including excreta on their own or mixed with water in the case of domestic wastewater (sewage), are:

- Fertilizer and irrigation water in agriculture and horticulture: for example using recovered and treated water for irrigation; using composted excreta (and other organic waste) or appropriately treated biosolids as fertilizer and soil conditioner; using treated source-separated urine as fertilizer.

- Energy: for example digesting feces and other organic waste to produce biogas; or producing combustible fuels.

- Other: other emerging excreta reuse options include producing protein feeds for livestock using black soldier fly larvae, recovering organic matter for use as building material or in paper production.

Multiple Barrier Concept for Safe use in Agriculture

Research into how to make reuse of urine and feces safe in agriculture was carried out in Sweden since the 1990s. In 2006 the World Health Organization (WHO) provided guidelines on safe reuse of wastewater, excreta and greywater. The multiple barrier concept to reuse, which is the key cornerstone of this publication, has led to a clear understanding on how excreta reuse can be done safely. The concept is also used in water supply and food production and is generally understood as a series of treatment steps and other safety precautions to prevent the spread of pathogens.

The degree of treatment required for excreta-based fertilizers before they can safely be used in agriculture depends on a number of factors. It mainly depends on which other barriers will be put in place according to the multiple barrier concept. Such barriers might be selecting a suitable crop, farming methods, methods of applying the fertilizer, education and so forth.

For example, in the case of urine-diverting dry toilets (UDDTs) secondary treatment of dried feces can be performed at community level rather than at household level and can include thermophilic composting where fecal material is composted at over 50 °C, prolonged storage with the duration of 1.5 to two years, chemical treatment with ammonia from urine to inactivate the pathogens, solar sanitation for further drying or heat treatment to eliminate pathogens.

Comparison of Excreta-based Fertilizers to Other Fertilizers

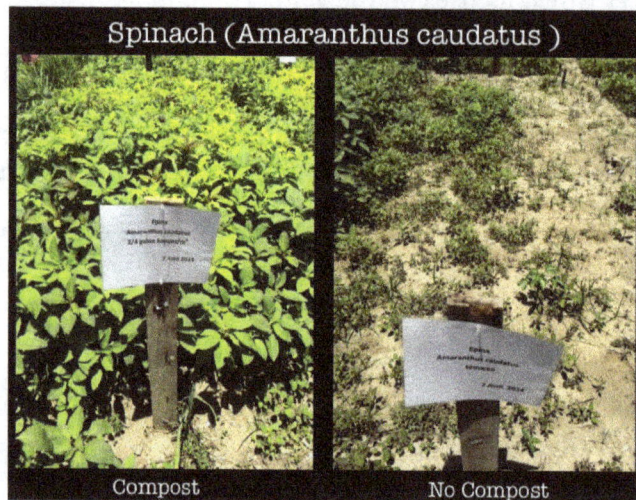

Comparison of spinach field with (left) and without (right) compost, experiments at the SOIL farm in Port-au-Prince, Haiti

There is an untapped fertilizer resource in human excreta. In Africa, for example, the theoretical quantities of nutrients that can be recovered from human excreta are comparable with all current fertilizer use on the continent. Reuse can therefore support increased food production and also provide an alternative to chemical fertilizers, which is often unaffordable to small-holder farmers.

Mineral fertilizers are made from mining activities and can contain heavy metals. Phosphate ores contain heavy metals such as cadmium and uranium, which can reach the food chain via mineral phosphate fertilizer. This does not apply to excreta-based fertilizers which is an advantage.

In intensive agricultural land use, animal manure is often not used as targeted as mineral fertilizers and thus the nitrogen utilization efficiency is poor. Animal manure can become a problem in terms of excessive use in areas of intensive agriculture with high numbers of livestock and too little available farmland.

Fertilizing elements of organic fertilizers are mostly bound in carbonaceous reduced compounds. If these are already partially oxidized as in the compost, the fertilizing minerals are adsorbed on the degradation products (humic acids) etc. Thus, they exhibit a slow-release effect and are usually less rapidly leached compared to mineral fertilizers.

Peak Phosphorus

In the case of phosphorus in particular, reuse of excreta is one known method to recover phosphorus to mitigate the looming shortage (also known as "peak phosphorus") of economical mined phosphorus. Mined phosphorus is a limited resource that is being used up for fertilizer production at an ever-increasing rate, which is threatening worldwide food security. Therefore, phosphorus from excreta-based fertilizers is an interesting alternative to fertilizers containing mined phosphate ore.

Uses as Excreta-derived Fertilizers

Urine

Application of urine on a field near Bonn, Germany, by means of flexible hose close to the soil

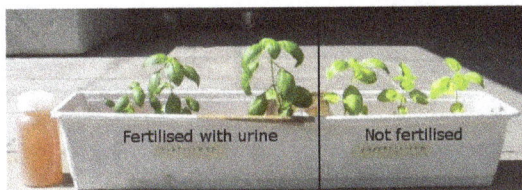
Basil plants: The plants on the right are not fertilized, while the plants on the left are fertilized with urine - in a nutrient-poor soil

Application of urine on eggplants during a comprehensive urine application field testing study at Xavier University, Philippines

Urine is primarily composed of water and urea. It contains large quantities of nitrogen (mostly as urea), as well as reasonable quantities of dissolved potassium. The nutrient concentrations in urine vary with diet. In particular nitrogen content in urine is related to quantity of protein in the diet. A high protein diet results in high urea levels in urine. Urine typically contains 70% of the nitrogen and more than half the potassium found in sewage, while making up less than 1% of the overall volume.

Urine fertilizer is usually applied diluted with water because undiluted urine can chemically burn the leaves or roots of some plants, particularly if the soil moisture content is low. The dilution also helps to reduce odour development following application. Applying urine as fertilizer has been called "closing the cycle of agricultural nutrient flows" or ecological sanitation or ecosan.

When diluted with water (at a 1:5 ratio for container-grown annual crops with fresh growing medium each season or a 1:8 ratio for more general use), it can be applied directly to soil as a fertilizer. The fertilization effect of urine has been found to be comparable to that of commercial nitrogen fertilizers. Concentrations of heavy metals such as lead, mercury, and cadmium, commonly found in sewage sludge, which is also used as fertilizer, are much lower in urine. Urine may contain pharmaceutical residues (environmental persistent pharmaceutical pollutants).

The more general limitations to using urine as fertilizer then depend mainly on the potential for buildup of excess nitrogen (due to the high ratio of that macronutrient), and inorganic salts such as sodium chloride, which are also part of the wastes excreted by the renal system. Over-fertilization with urine or other nitrogen fertilizers can result in too much ammonia for plants to absorb, acidic conditions, or other phytotoxicity. The degree to which these factors impact the effectiveness depends on the term of use, salinity tolerance of the plant, soil composition, addition of other fertilizing compounds, and quantity of rainfall or other irrigation.

Human urine can be collected with sanitation systems that utilize urinals or urine diversion toilets. If urine is to be collected for use as a fertilizer in agriculture, then the easiest method of doing so is (in increasing order of costs) by using waterless urinals, urine-diverting dry toilets (UDDTs) or urine diversion flush toilets.

The risks of using urine as a natural source of agricultural fertilizer are generally regarded as negligible or acceptable. There are potentially more environmental problems (such as eutrophication resulting from the influx of nutrient rich effluent into aquatic or marine ecosystems) and a higher energy consumption when urine is treated as part of sewage in wastewater treatment plants compared with when it is used directly as a fertilizer resource.

In developing countries the use of raw sewage or fecal sludge has been common throughout history, yet the application of pure urine to crops is rare. Increasingly there are calls for urine's use as a fertilizer.

Since about 2011, the Bill and Melinda Gates Foundation is providing research funding sanitation systems that recover the nutrients in urine.

Examples

- In Tororo District in eastern Uganda - a region with severe land degradation problems - smallholder farmers appreciated urine fertilization as a low-cost and low-risk practice which can contribute to significant yield increases. The importance of social norms and cultural perceptions needs to be recognized but are not absolute barriers to adoption of the practice.

Dried Feces

Reuse of dried feces (feces) from urine-diverting dry toilets (UDDTs) after post-treatment can result in increased crop production through fertilizing effects of nitrogen, phosphorus, potassium and improved soil fertility through organic carbon.

Composted Feces

Cabbage grown in excreta-based compost (left) and without soil amendments (right), SOIL in Haiti

Compost derived from composting toilets (where organic kitchen waste is in some cases also added to the composting toilet), has in principal the same uses as compost derived from other organic waste products, such as sewage sludge or municipal organic waste. One limiting factor may be legal restrictions due to the possibility that pathogens remain in the compost. In any case, the use of compost from composting toilets in one's own garden can be regarded as safe and is the main method of use for compost from composting toilets. Hygienic measures for handling of the compost must be applied by all those people who are exposed to it, e.g. wearing gloves and boots.

Some of the urine will be part of the compost, although some urine will be lost via leachate and evaporation. Urine can contain up to 90 percent of the nitrogen, up to 50 percent of the phosphorus, and up to 70 percent of the potassium present in human excreta.

The nutrients in compost from a composting toilet have a higher plant availability than the product (dried feces) from a urine-diverting dry toilet.

Fecal Sludge

Fecal sludge (also called septage) is defined as "coming from onsite sanitation technologies, and has not been transported through a sewer." Examples of onsite technologies include pit latrines, unsewered public ablution blocks, septic tanks and dry toilets. Fecal sludge can be treated by a variety of methods to render it suitable for reuse in agriculture. These include (usually carried out in combination): dewatering, thickening, drying (in sludge drying beds), composting, pelletization, anaerobic digestion.

Municipal Wastewater

Reclaimed water can be reused for irrigation, industrial uses, replenishing natural water courses, water bodies and aquifers and other potable and non-potable uses. These applications however focus usually on the water aspect, not on the nutrients and organic matter reuse aspect, which is the focus of "reuse of excreta".

When wastewater is reused in agriculture, its nutrient (nitrogen and phosphorus) content may be useful for additional fertilizer application. Work by the International Water Management Institute and others has led to guidelines on how reuse of municipal wastewater in agriculture for irrigation and fertilizer application can be safely implemented in low income countries.

Sewage Sludge

The use of treated sewage sludge (after treatment also called "biosolids") as a soil conditioner or fertilizer is possible but is a controversial topic in some countries (such as USA, some countries in Europe) due to the chemical pollutants it may contain, such as heavy metals and environmental persistent pharmaceutical pollutants.

Northumbrian Water in the United Kingdom uses two biogas plants to produce what the company calls "poo power" - using sewage sludge to produce energy to generate income. Biogas production has reduced its pre 1996 electricity expenditure of 20 million GBP by about 20%. Severn Trent and Wessex Water also have similar projects.

Sludge Treatment Liquids

Sludge treatment liquids (after anaerobic digestion) can be used as an input source for a process to recover phosphorus in the form of struvite for use as fertilizer. For example, the Canadian company Ostara Nutrient Recovery Technologies is marketing a process based on controlled chemical precipitation of phosphorus in a fluidized bed reactor that recovers struvite in the form of crystalline pellets from sludge dewatering streams. The resulting crystalline product is sold to the agriculture, turf and ornamental plants sectors as fertilizer under the registered trade name "Crystal Green".

Animal Manure

Animal dung (manure) has been used for centuries as a fertilizer for farming, as it improves the soil structure (aggregation), so that it holds more nutrients and water, and becomes more fertile. Animal manure also encourages soil microbial activity, which promotes the soil's trace mineral supply, improving plant nutrition. It also contains some nitrogen and other nutrients that assist the growth of plants.

Manures with a particularly unpleasant odor (such as slurry from intensive pig farming) are usually knifed (injected) directly into the soil to reduce release of the odor. Manure from pigs and cattle is usually spread on fields using a manure spreader. Due to the relatively lower level of proteins in vegetable matter, herbivore manure has a milder smell than the dung of carnivores or omnivores. However, herbivore slurry that has undergone anaerobic fermentation may develop more unpleasant odors, and this can be a problem in some agricultural regions. Poultry droppings are harmful to plants when fresh but, after a period of composting, are valuable fertilizers.

Manure is also commercially composted and bagged and sold retail as a soil amendment.

Health and Environmental Aspects of Agricultural uses

Pathogens

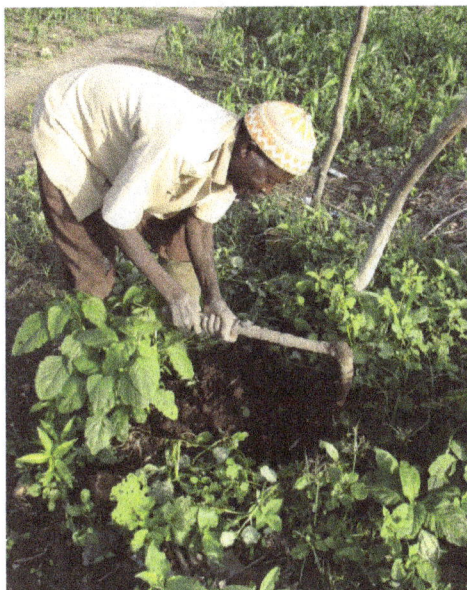

Gardeners of Fada N'Gourma in Burkina Faso apply dry excreta after mixing with other organic fertilizer (donkey manure, cow manure) and pure fertile soil, and after maturing for another 2 to 4 months

Risk Management

Exposure of farm workers to untreated excreta constitutes a significant health risk due to its pathogen content. This risk also extends to consumers of crops fertilized with untreated excreta. Therefore, excreta need to be appropriately treated before reuse, and health aspects need to be managed for all reuse applications as the excreta can contain pathogens even after treatment.

Treatment of Excreta for Pathogen Removal

The treatment of excreta and wastewater for pathogen removal can take place:

- at the toilet itself (for example urine collected from urine-diverting dry toilets is often treated by simple storage at the household level); or

- at a semi-centralized level for example by composting; or

- at a fully centralized level at sewage treatment plants and sewage sludge treatment plants.

Indicator Organisms

As an indicator organism in reuse schemes, helminth eggs are commonly used as these organisms are the most difficult to destroy in most treatment processes. The multiple barrier approach is recommended where e.g. lower levels of treatment may be acceptable when combined with other post-treatment barriers along the sanitation chain.

Pharmaceutical Residues

Excreta from humans and farmed animals contain hormones and pharmaceutical residues which could in theory enter the food chain via fertilized crops but are currently not fully removed by conventional wastewater treatment plants anyway and can enter drinking water sources via household wastewater (sewage). In fact, the pharmaceutical residues in the excreta are degraded better in terrestrial systems (soil) than in aquatic systems.

Nitrate Pollution

Only a fraction of the nitrogen-based fertilizers is converted to produce and other plant matter. The remainder accumulates in the soil or lost as run-off. This also applies to excreta-based fertilizer since it also contains nitrogen. Excessive nitrogen which is not taken up by plants is transformed into nitrate which is easily leached. High application rates combined with the high water-solubility of nitrate leads to increased runoff into surface water as well as leaching into groundwater. Nitrate levels above 10 mg/L (10 ppm) in groundwater can cause 'blue baby syndrome' (acquired methemoglobinemia). The nutrients, especially nitrates, in fertilizers can cause problems for natural habitats and for human health if they are washed off soil into watercourses or leached through soil into groundwater.

Society and Culture

Policies

There is still a lack of examples of implemented policy where the reuse aspect is fully integrated in policy and advocacy. Revising legislation does not necessarily lead to functioning reuse systems; it is important to describe the "institutional landscape" and involve all actors; parallel processes should be initiated at all levels of government i.e. national, regional and local level); country specific strategies and approaches are needed; and strategies supporting newly developed policies need to be developed.

Regulations

In countries that export agricultural products, excreta reuse as a means to reuse nutrients may be hindered by regulations in countries to which they export.

Urine use in Organic Farming in Europe

The European Union (EU) only allows the use of source separated urine in conventional farming within the EU, but not yet in organic farming. This is a situation that many agricultural experts, especially in Sweden, would like to see changed. This ban may also reduce the options to use urine as a fertilizer in other countries if they wish to export their products to the EU.

Dried Feces from UDDTs in the U.S.

In the United States, the EPA regulation governs the management of sewage sludge but has no jurisdiction over the byproducts of a urine-diversion dry toilet (UDDT). Oversight of these materials falls to the states.

Examples

Haiti

The NGO SOIL in Haiti began building urine-diverting dry toilets and composting the waste produced for agricultural use in Haiti in 2006. SOIL's two composting waste treatment facilities currently transform over 20,000 gallons (75,708 liters) of human excreta into safe, organic, agricultural-grade compost every month. The compost produced at these facilities is sold to farmers, organizations, businesses, and institutions around the country to help finance SOIL's waste treatment operations. Crops grown with this soil amendment include spinach, peppers, sorghum, maize, and more. Each batch of compost produced is tested for the indicator organism *E. coli* to ensure that complete pathogen kill has taken place during the thermophilic composting process.

History of Agricultural Reuse

The recovery and use of excreta, with or without mixing with water, has been practiced by almost all cultures over the millennia.

Other uses of Excreta (non fertilizer)

Apart from use in agriculture, there are other possible uses of excreta. For example, in the case of fecal sludge, it can be treated and then serve as protein (black soldier fly process), fodder, fish food, building materials and biofuels (biogas from anaerobic digestion, incineration or co-combustion of dried sludge, pyrolysis of fecal sludge, biodiesel from fecal sludge).

Fuel

Dry animal dung is used as a fuel in many countries around the world.

Pilot scale research in Uganda and Senegal has shown that it is viable to use dry feces as for combustion in industry, provided it has been dried to a minimum of 28% dry solids.

Biogas

Small-scale biogas plants exist in several countries, e.g. Ghana, utilizing mixed animal and human feces, and larger centralized systems are being planned. The same use of excreta - when it is mixed with domestic wastewater - takes place in the anaerobic digesters of centralized sewage treatment plants where sewage sludge is digested and biogas is produced. This biogas is used for heating the digesters and for electricity generation.

Production of Animal Protein

Pilot facilities are being developed for feeding Black Soldier Fly larvae with feces to produce protein as a feed for chickens in South Africa.

Building Materials

It is known that additions of fecal matter up to 20% by dried weight in clay bricks does not make a significant functional difference to bricks.

References

- Kaya, C; Kirnak, H; Higgs, D; Saltali, K (2002-02-28). "Supplementary calcium enhances plant growth and fruit yield in strawberry cultivars grown at high (NaCl) salinity". Scientia Horticulturae. 93 (1): 65–74. doi:10.1016/S0304-4238(01)00313-2

- Kopittke, Peter M.; Menzies, Neal W.; Wang, Peng; Blamey, F. Pax C. (August 2016). "Kinetics and nature of aluminium rhizotoxic effects: a review". Journal of Experimental Botany. 67 (15): 4451–4467. doi:10.1093/jxb/erw233

- Marschner, Petra, ed. (2012). Marschner's mineral nutrition of higher plants (3rd ed.). Amsterdam: Elsevier/Academic Press. ISBN 9780123849052

- "AAPFCO Board of Directors 2006 Mid-Year Meeting" (PDF). Association of American Plant Food Control Officials. Retrieved 18 July 2011

- van Breemen, N.; Mulder, J.; Driscoll, C. T. (October 1983). "Acidification and alkalinization of soils". Plant and Soil. 75 (3): 283–308. doi:10.1007/BF02369968

- Shavrukov, Yuri; Hirai, Yoshihiko (January 2016). "Good and bad protons: genetic aspects of acidity stress responses in plants". Journal of Experimental Botany. 67 (1): 15–30. doi:10.1093/jxb/erv437

- Bloom, Paul R.; Skyllberg, Ulf (2012). "Soil pH and pH buffering". In Huang, Pan Ming; Li, Yuncong; Sumner, Malcolm E. Handbook of soil sciences : properties and processes (2nd ed.). Boca Raton, FL: CRC Press. pp. 19–1 to 19–14. ISBN 9781439803059

- White, Philip J. (2016). "Selenium accumulation by plants". Annals of Botany. 117: 217–235. doi:10.1093/aob/mcv180. Retrieved 5 June 2016

- Van Breemen, N.; Driscoll, C. T.; Mulder, J. (16 February 1984). "Acidic deposition and internal proton sources in acidification of soils and waters". Nature. 307 (5952): 599–604. doi:10.1038/307599a0

- "Journal of Contaminant Hydrology - Fertilizer-N use efficiency and nitrate pollution of groundwater in developing countries". Journal of Contaminant Hydrology. 20: 167–184. doi:10.1016/0169-7722(95)00067-4. Retrieved 17 June 2012

- Steinfeld, Carol (2004). Liquid Gold: The Lore and Logic of Using Urine to Grow Plants. Ecowaters Books. ISBN 978-0-9666783-1-4

- Soil Survey Division Staff. "Soil survey manual. 1993. Chapter 3.". Soil Conservation Service. U.S. Department of Agriculture Handbook 18. Retrieved 2017-05-15

- Sumner, Malcolm E.; Yamada, Tsuioshi (November 2002). "Farming with acidity". Communications in Soil Science and Plant Analysis. 33 (15–18): 2467–2496. doi:10.1081/CSS-120014461

- Andersson, Elina (June 1, 2015). "Turning waste into value: using human urine to enrich soils for sustainable food production in Uganda". Journal of Cleaner Production. Integrating Cleaner Production into Sustainability Strategies. 96: 290–298. doi:10.1016/j.jclepro.2014.01.070

- "Land Application of Municipal Biosolids". Environmental Health - Toxic Substances. United States Geological Survey. Retrieved 24 April 2015

- Linda Strande, Mariska Ronteltap, Damir Brdjanovic (eds.) (2013). Faecal sludge management : systems approach for implementation and operation. IWA Publishing. ISBN 9781780404721

- EPA 832-F-99-066, September 1999. "Water Efficiency Technology Fact Sheet Composting Toilets" (PDF). United States Environmental Protection Agency. Office of Water. Retrieved 3 January 2015

Production Processes of Nitrogen Fertilizers

Nitrogen fertilizers are produced from ammonia and can be injected into the ground directly. Ammonia is a colorless gas which is a compound of hydrogen and nitrogen. The diverse applications of nitrogen fertilizers in the current scenario have been thoroughly discussed in this chapter.

Ammonia

Ammonia or azane is a compound of nitrogen and hydrogen with the formula NH_3. The simplest pnictogen hydride, ammonia is a colourless gas with a characteristic pungent smell. It contributes significantly to the nutritional needs of terrestrial organisms by serving as a precursor to food and fertilizers. Ammonia, either directly or indirectly, is also a building block for the synthesis of many pharmaceutical products and is used in many commercial cleaning products.

Although common in nature and in wide use, ammonia is both caustic and hazardous in its concentrated form. It is classified as an extremely hazardous substance in the United States as defined in Section 302 of the U.S. Emergency Planning and Community Right-to-Know Act (42 U.S.C. 11002), and is subject to strict reporting requirements by facilities which produce, store, or use it in significant quantities.

The global industrial production of ammonia in 2014 was 176 million tonnes (173,000,000 long tons; 194,000,000 short tons), a 16% increase over the 2006 global industrial production of 152 million tonnes (150,000,000 long tons; 168,000,000 short tons). Industrial ammonia is sold either as ammonia liquor (usually 28% ammonia in water) or as pressurized or refrigerated anhydrous liquid ammonia transported in tank cars or cylinders.

NH_3 boils at −33.34 °C (−28.012 °F) at a pressure of one atmosphere, so the liquid must be stored under pressure or at low temperature. Household ammonia or ammonium hydroxide is a solution of NH_3 in water. The concentration of such solutions is measured in units of the Baumé scale (density), with 26 degrees baumé (about 30% (by weight) ammonia at 15.5 °C or 59.9 °F) being the typical high-concentration commercial product.

Natural Occurrence

Ammonia is found in trace quantities in nature, being produced from the nitrogenous animal and vegetable matter. Ammonia and ammonium salts are also found in small quantities in rainwater, whereas ammonium chloride (sal ammoniac), and ammonium sulfate are found in volcanic districts; crystals of ammonium bicarbonate have been found in Patagonian guano. The kidneys secrete ammonia to neutralize excess acid. Ammonium salts are found distributed through fertile soil and in seawater.

Ammonia is also found throughout the Solar System on Mars, Jupiter, Saturn, Uranus, Neptune, and Pluto, among other places: on smaller, icy worlds like Pluto, ammonia can act as a geologically important antifreeze, as a mixture of water and ammonia can potentially have a melting point of as low as 173 kelvins if the ammonia concentration is high enough and thus allow such worlds to retain internal oceans and active geology far longer than would be possible with water alone. Substances containing ammonia, or those that are similar to it, are called ammoniacal.

Properties

Ammonia is a colourless gas with a characteristic pungent smell. It is lighter than air, its density being 0.589 times that of air. It is easily liquefied due to the strong hydrogen bonding between molecules; the liquid boils at −33.3 °C (−27.94 °F), and freezes at −77.7 °C (−107.86 °F) to white crystals.

Ammonia may be conveniently deodorized by reacting it with either sodium bicarbonate or acetic acid. Both of these reactions form an odourless ammonium salt.

Solid

The crystal symmetry is cubic, Pearson symbol cP16, space group $P2_13$ No.198, lattice constant 0.5125 nm.

Liquid

Liquid ammonia possesses strong ionising powers reflecting its high ε of 22. Liquid ammonia has a very high standard enthalpy change of vaporization (23.35 kJ/mol, cf. water 40.65 kJ/mol, methane 8.19 kJ/mol, phosphine 14.6 kJ/mol) and can therefore be used in laboratories in uninsulated vessels without additional refrigeration.

Solvent Properties

Ammonia is miscible with water. In an aqueous solution, it can be expelled by boiling. The aqueous solution of ammonia is basic. The maximum concentration of ammonia in water (a saturated solution) has a density of 0.880 g/cm³ and is often known as '.880 ammonia'. Ammonia does not burn readily or sustain combustion, except under narrow fuel-to-air mixtures of 15–25% air.

Combustion

When mixed with oxygen, it burns with a pale yellowish-green flame. At high temperature and in the presence of a suitable catalyst, ammonia is decomposed into its constituent elements. Ignition occurs when chlorine is passed into ammonia, forming nitrogen and hydrogen chloride; if chlorine is present in excess, then the highly explosive nitrogen trichloride (NCl_3) is also formed.

Structure

The ammonia molecule has a trigonal pyramidal shape as predicted by the valence shell electron pair repulsion theory (VSEPR theory) with an experimentally determined bond angle of $106.7°$. The central nitrogen atom has five outer electrons with an additional electron from each hydrogen atom. This gives a total of eight electrons, or four electron pairs that are arranged tetrahedrally. Three of these electron pairs are used as bond pairs, which leaves one lone pair of electrons. The lone pair of electrons repel more strongly than bond pairs, therefore the bond angle is not $109.5°$, as expected for a regular tetrahedral arrangement, but $106.7°$. The nitrogen atom in the molecule has a lone electron pair, which makes ammonia a base, a proton acceptor. This shape gives the molecule a dipole moment and makes it polar. The molecule's polarity and, especially, its ability to form hydrogen bonds, makes ammonia highly miscible with water. Ammonia is moderately basic, a 1.0 M aqueous solution has a pH of 11.6 and if a strong acid is added to such a solution until the solution is neutral (pH = 7), 99.4% of the ammonia molecules are protonated. Temperature and salinity also affect the proportion of NH_4^+. The latter has the shape of a regular tetrahedron and is isoelectronic with methane.

The ammonia molecule readily undergoes nitrogen inversion at room temperature; a useful analogy is an umbrella turning itself inside out in a strong wind. The energy barrier to this inversion is 24.7 kJ/mol, and the resonance frequency is 23.79 GHz, corresponding to microwave radiation of a wavelength of 1.260 cm. The absorption at this frequency was the first microwave spectrum to be observed.

Amphotericity

One of the most characteristic properties of ammonia is its basicity. Ammonia is considered to be a weak base. It combines with acids to form salts; thus with hydrochloric acid it forms ammonium chloride (sal ammoniac); with nitric acid, ammonium nitrate, etc. Perfectly dry ammonia will not combine with perfectly dry hydrogen chloride; moisture is necessary to bring about the reaction. As a demonstration experiment, opened bottles of concentrated ammonia and hydrochloric acid produce clouds of ammonium chloride, which seem to appear "out of nothing" as the salt forms where the two diffusing clouds of molecules meet, somewhere between the two bottles.

$$NH_3 + HCl \rightarrow NH_4Cl$$

The salts produced by the action of ammonia on acids are known as the ammonium salts and all contain the ammonium ion (NH_4^+).

Although ammonia is well known as a weak base, it can also act as an extremely weak acid. It is a protic substance and is capable of formation of amides (which contain the NH_2^- ion). For example, lithium dissolves in liquid ammonia to give a solution of lithium amide:

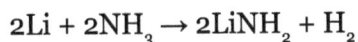

$$2Li + 2NH_3 \rightarrow 2LiNH_2 + H_2$$

Self-dissociation

Like water, ammonia undergoes molecular autoionisation to form its acid and base conjugates:

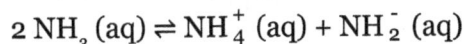

$$2\,NH_3\,(aq) \rightleftharpoons NH_4^+\,(aq) + NH_2^-\,(aq)$$

Ammonia often functions as a weak base, so it has some buffering ability. Shifts in pH will cause more or fewer ammonium cations (NH_4^+) and amide anions (NH_2^-) to be present in solution. At standard pressure and temperature, $K=[NH_4^+][NH_2^-] = 10^{-30}$

Combustion

The combustion of ammonia to nitrogen and water is exothermic:

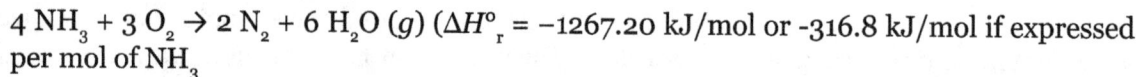

$$4\,NH_3 + 3\,O_2 \rightarrow 2\,N_2 + 6\,H_2O\ (g)\ (\Delta H^\circ_r = -1267.20\ \text{kJ/mol or } -316.8\ \text{kJ/mol if expressed}$$
per mol of NH_3

The standard enthalpy change of combustion, ΔH°_c, expressed per mole of ammonia and with condensation of the water formed, is −382.81 kJ/mol. Dinitrogen is the thermodynamic product of combustion: all nitrogen oxides are unstable with respect to N_2 and O_2, which is the principle behind the catalytic converter. Nitrogen oxides can be formed as kinetic products in the presence of appropriate catalysts, a reaction of great industrial importance in the production of nitric acid:

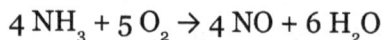

$$4\,NH_3 + 5\,O_2 \rightarrow 4\,NO + 6\,H_2O$$

A subsequent reaction leads to NO_2

$$2\,NO + O_2 \rightarrow 2\,NO_2$$

The combustion of ammonia in air is very difficult in the absence of a catalyst (such as platinum gauze or warm chromium(III) oxide), because the temperature of the flame is usually lower than the ignition temperature of the ammonia−air mixture. The flammable range of ammonia in air is 16−25%.

Formation of Other Compounds

In organic chemistry, ammonia can act as a nucleophile in substitution reactions. Amines can be formed by the reaction of ammonia with alkyl halides, although the resulting -NH_2 group is also nucleophilic and secondary and tertiary amines are often formed as byproducts. An excess of ammonia helps minimise multiple substitution, and neutralises the hydrogen halide formed. Methylamine is prepared commercially by the reaction of ammonia with chloromethane, and the reaction of ammonia with 2-bromopropanoic acid has been used to prepare racemic alanine in 70% yield. Ethanolamine is prepared by a ring-opening reaction with ethylene oxide: the reaction is sometimes allowed to go further to produce diethanolamine and triethanolamine.

Amides can be prepared by the reaction of ammonia with carboxylic acid derivatives. Acyl chlorides are the most reactive, but the ammonia must be present in at least a twofold excess to neutralise the hydrogen chloride formed. Esters and anhydrides also react with ammonia to form amides. Ammonium salts of carboxylic acids can be dehydrated to amides so long as there are no thermally sensitive groups present: temperatures of 150−200 °C are required.

The hydrogen in ammonia is capable of replacement by metals, thus magnesium burns in the gas with the formation of magnesium nitride Mg_3N_2, and when the gas is passed over heated sodium or potassium, sodamide, $NaNH_2$, and potassamide, KNH_2, are formed. Where necessary in substitutive nomenclature, IUPAC recommendations prefer the name "azane" to ammonia: hence chloramine would be named "chloroazane" in substitutive nomenclature, not "chloroammonia".

Pentavalent ammonia is known as λ^5-amine, or more commonly, ammonium hydride. This crystalline solid is only stable under high pressure, and decomposes back into trivalent ammonia and hydrogen gas at normal conditions. This substance was once investigated as a possible solid rocket fuel in 1966.

Ammonia as a Ligand

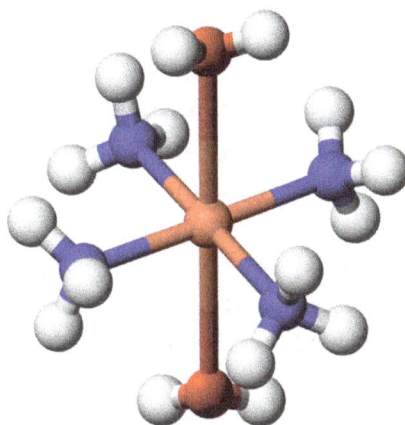

Ball-and-stick model of the tetraamminediaquacopper(II) cation, $[Cu(NH_3)_4(H_2O)_2]^{2+}$

Ammonia can act as a ligand in transition metal complexes. It is a pure σ-donor, in the middle of the spectrochemical series, and shows intermediate hard-soft behaviour. For historical reasons, ammonia is named ammine in the nomenclature of coordination compounds. Some notable ammine complexes include tetraamminediaquacopper(II) ($[Cu(NH_3)_4(H_2O)_2]^{2+}$), a dark blue complex formed by adding ammonia to a solution of copper(II) salts. Tetraamminediaquacopper(II) hydroxide is known as Schweizer's reagent, and has the remarkable ability to dissolve cellulose. Diamminesilver(I) ($[Ag(NH_3)_2]^+$) is the active species in Tollens' reagent. Formation of this complex can also help to distinguish between precipitates of the different silver halides: silver chloride (AgCl) is soluble in dilute (2M) ammonia solution, silver bromide (AgBr) is only soluble in concentrated ammonia solution, whereas silver iodide (AgI) is insoluble in aqueous ammonia.

Ammine complexes of chromium(III) were known in the late 19th century, and formed the basis of Alfred Werner's revolutionary theory on the structure of coordination compounds. Werner noted only two isomers (fac- and mer-) of the complex $[CrCl_3(NH_3)_3]$ could be formed, and concluded the ligands must be arranged around the metal ion at the vertices of an octahedron. This proposal has since been confirmed by X-ray crystallography.

An ammine ligand bound to a metal ion is markedly more acidic than a free ammonia molecule, although deprotonation in aqueous solution is still rare. One example is the Calomel reaction, where the resulting amidomercury(II) compound is highly insoluble.

$$Hg_2Cl_2 + 2\,NH_3 \rightarrow Hg + HgCl(NH_2) + NH_4^+ + Cl^-$$

Detection and Determination

Ammonia in Solution

Ammonia and ammonium salts can be readily detected, in very minute traces, by the addition of Nessler's solution, which gives a distinct yellow colouration in the presence of the least trace of ammonia or ammonium salts. The amount of ammonia in ammonium salts can be estimated quantitatively by distillation of the salts with sodium or potassium hydroxide, the ammonia evolved being absorbed in a known volume of standard sulfuric acid and the excess of acid then determined volumetrically; or the ammonia may be absorbed in hydrochloric acid and the ammonium chloride so formed precipitated as ammonium hexachloroplatinate, $(NH_4)_2PtCl_6$.

Gaseous Ammonia

Sulfur sticks are burnt to detect small leaks in industrial ammonia refrigeration systems. Larger quantities can be detected by warming the salts with a caustic alkali or with quicklime, when the characteristic smell of ammonia will be at once apparent. Ammonia is an irritant and irritation increases with concentration; the Permissible Exposure Limit is 25 ppm, and lethal above 500 ppm. Higher concentrations are hardly detected by conventional detectors, the type of detector is chosen according to the sensitivity required (e.g. semiconductor, catalytic, electrochemical). Holographic sensors have been proposed for detecting concentrations up to 12.5% in volume.

Ammoniacal Nitrogen ($NH_3^- N$)

Ammoniacal nitrogen ($NH_3^- N$) is a measure commonly used for testing the quantity of ammonium ions, derived naturally from ammonia, and returned to ammonia via organic processes, in water or waste liquids. It is a measure used mainly for quantifying values in waste treatment and water purification systems, as well as a measure of the health of natural and man made water reserves. It is measured in units of mg/L (milligram per litre).

History

This high-pressure reactor was built in 1921 by BASF in Ludwigshafen and was re-erected on the premises of the University of Karlsruhe in Germany.

The ancient Greek historian Herodotus mentioned that there were outcrops of salt in an area of Libya that was inhabited by a people called the "Ammonians" (now: the Siwa oasis in northwestern Egypt, where salt lakes still exist). The Greek geographer Strabo also mentioned the salt from this region. However, the ancient authors Dioscorides, Apicius, Arrian, Synesius, and Aëtius of Amida described this salt as forming clear crystals that could be used for cooking and that were essentially rock salt. *Hammoniacus sal* appears in the writings of Pliny, although it is not known whether the term is identical with the more modern sal ammoniac (ammonium chloride).

The fermentation of urine by bacteria produces a solution of ammonia; hence fermented urine was used in Classical Antiquity to wash cloth and clothing, to remove hair from hides in preparation for tanning, to serve as a mordant in dying cloth, and to remove rust from iron.

In the form of sal ammoniac ammonia was important to the Muslim alchemists as early as the 8th century, first mentioned by the Persian chemist Jābir ibn Hayyān, and to the European alchemists since the 13th century, being mentioned by Albertus Magnus. It was also used by dyers in the Middle Ages in the form of fermented urine to alter the colour of vegetable dyes. In the 15th century, Basilius Valentinus showed that ammonia could be obtained by the action of alkalis on sal ammoniac. At a later period, when sal ammoniac was obtained by distilling the hooves and horns of oxen and neutralizing the resulting carbonate with hydrochloric acid, the name "spirit of hartshorn" was applied to ammonia.

Gaseous ammonia was first isolated by Joseph Black in 1756 by reacting *sal ammoniac* (Ammonium Chloride) with *calcined magnesia* (Magnesium Oxide). It was isolated again by Peter Woulfe in 1767 and by Joseph Priestley in 1773 and was termed by him "alkaline air". Eleven years later in 1785, Claude Louis Berthollet ascertained its composition.

The Haber–Bosch process to produce ammonia from the nitrogen in the air was developed by Fritz Haber and Carl Bosch in 1909 and patented in 1910. It was first used on an industrial scale in Germany during World War I, following the allied blockade that cut off the supply of nitrates from Chile. The ammonia was used to produce explosives to sustain war efforts.

Prior to the availability of natural gas, hydrogen as a precursor to ammonia production was produced via the electrolysis of water or using the chloralkali process.

With the advent of the steel industry in the 20th century, ammonia became a byproduct of the production of coking coal.

Uses

Fertilizer

Globally, approximately 88% (as of 2014) of ammonia is used as fertilizers either as its salts, solutions or anhydrously. When applied to soil, it helps provide increased yields of crops such as maize and wheat. 30% of agricultural nitrogen applied in the USA is in the form of anhydrous ammonia and worldwide 110 million tonnes are applied each year.

Precursor to Nitrogenous Compounds

Ammonia is directly or indirectly the precursor to most nitrogen-containing compounds. Virtually all synthetic nitrogen compounds are derived from ammonia. An important derivative is nitric acid. This key material is generated via the Ostwald process by oxidation of ammonia with air over a platinum catalyst at 700–850 °C (1,292–1,562 °F), ~9 atm. Nitric oxide is an intermediate in this conversion:

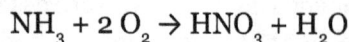

$$NH_3 + 2 O_2 \rightarrow HNO_3 + H_2O$$

Nitric acid is used for the production of fertilizers, explosives, and many organonitrogen compounds.

Ammonia is also used to make the following compounds:

- Hydrazine, in the Olin Raschig process and the peroxide process

- Hydrogen cyanide, in the BMA process and the Andrussow process

- Hydroxylamine and ammonium carbonate, in the Raschig process

- Phenol, in the Raschig–Hooker process

- Urea, in the Bosch–Meiser urea process and in Wöhler synthesis

- Amino acids, using Strecker amino-acid synthesis

- Acrylonitrile, in the Sohio process

Ammonia can also be used to make compounds in reactions which are not specifically named. Examples of such compounds include: ammonium perchlorate, ammonium nitrate, formamide, dinitrogen tetroxide, alprazolam, ethanolamine, ethyl carbamate, hexamethylenetetramine, and ammonium bicarbonate.

Cleaner

Household ammonia is a solution of NH_3 in water (i.e., ammonium hydroxide) used as a general purpose cleaner for many surfaces. Because ammonia results in a relatively streak-free shine, one of its most common uses is to clean glass, porcelain and stainless steel. It is also frequently used for cleaning ovens and soaking items to loosen baked-on grime. Household ammonia ranges in concentration by weight from 5 to 10% ammonia.

Fermentation

Solutions of ammonia ranging from 16% to 25% are used in the fermentation industry as a source of nitrogen for microorganisms and to adjust pH during fermentation.

Antimicrobial Agent for Food Products

As early as in 1895, it was known that ammonia was "strongly antiseptic ... it requires 1.4 grams per litre to preserve beef tea." In one study, anhydrous ammonia destroyed 99.999% of zoonotic bacteria in 3 types of animal feed, but not silage. Anhydrous ammonia is currently used commercially to reduce or eliminate microbial contamination of beef. Lean finely textured beef in the beef industry is made from fatty beef trimmings (c. 50–70% fat) by removing the fat using heat and centrifugation, then treating it with ammonia to kill *E. coli*. The process was deemed effective and safe by the US Department of Agriculture based on a study that found that the treatment reduces *E. coli* to undetectable levels. There have been safety concerns about the process as well as consumer complaints about the taste and smell of beef treated at optimal levels of ammonia. The level of ammonia in any final product has not come close to toxic levels to humans.

Minor and Emerging uses

Refrigeration – R717

Because of ammonia's vaporization properties, it is a useful refrigerant. It was commonly used prior to the popularisation of chlorofluorocarbons (Freons). Anhydrous ammonia is widely used in industrial refrigeration applications and hockey rinks because of its high energy efficiency and low cost. It suffers from the disadvantage of toxicity, which restricts its domestic and small-scale use. Along with its use in modern vapor-compression refrigeration it is used in a mixture along with hydrogen and water in absorption refrigerators. The Kalina cycle, which is of growing importance to geothermal power plants, depends on the wide boiling range of the ammonia–water mixture. Ammonia coolant is also used in the S1 radiator aboard the International Space Station in two loops which are used to regulate the internal temperature and enable temperature dependent experiments.

For Remediation of Gaseous Emissions

Ammonia is used to scrub SO_2 from the burning of fossil fuels, and the resulting product is converted to ammonium sulfate for use as fertilizer. Ammonia neutralizes the nitrogen oxides (NO_x) pollutants emitted by diesel engines. This technology, called SCR (selective catalytic reduction), relies on a vanadia-based catalyst.

Ammonia may be used to mitigate gaseous spills of phosgene.

As a Fuel

Ammoniacal Gas Engine Streetcar in New Orleans drawn by Alfred Waud in 1871.

The X-15 aircraft used ammonia as one component fuel of its rocket engine

The raw energy density of liquid ammonia is 11.5 MJ/L, which is about a third that of diesel. Although it can be used as a fuel, for a number of reasons this has never been common or widespread.

Ammonia engines or ammonia motors, using ammonia as a working fluid, have been proposed and occasionally used. The principle is similar to that used in a fireless locomotive, but with ammonia as the working fluid, instead of steam or compressed air. Ammonia engines were used experimentally in the 19th century by Goldsworthy Gurney in the UK and the St. Charles Avenue Streetcar line in New Orleans in the 1870s and 1880s, and during World War II ammonia was used to power buses in Belgium.

Ammonia is sometimes proposed as a practical alternative to fossil fuel for internal combustion engines. Its high octane rating of 120 and low flame temperature allows the use of high compression ratios without a penalty of high NOx production. Since ammonia contains no carbon, its combustion cannot produce carbon monoxide, hydrocarbons or soot.

However ammonia cannot be easily used in existing Otto cycle engines because of its very narrow flammability range and there are also other barriers to widespread automobile usage. In terms of raw ammonia supplies, plants would have to be built to increase production levels, requiring significant capital and energy sources. Although it is the second most produced chemical, the scale of ammonia production is a small fraction of world petroleum usage. It could be manufactured from renewable energy sources, as well as coal or nuclear power. The 60 MW Rjukan dam in Telemark, Norway produced ammonia for many years from 1913 producing fertilizer for much of Europe.

Despite this, several tests have been done. In 1981, a Canadian company converted a 1981 Chevrolet Impala to operate using ammonia as fuel. In 2007, a University of Michigan pickup powered by ammonia drove from Detroit to San Francisco as part of a demonstration, requiring only one fill-up in Wyoming.

Compared to hydrogen as a fuel, ammonia is much more energy efficient, and it would be a much lower cost to produce, store, and deliver hydrogen as ammonia than as compressed and/or cryogenic hydrogen. The conversion of ammonia to hydrogen via the sodium-amide process, either as a catalyst for combustion or as fuel for a proton exchange membrane fuel cell, is another possibility. Conversion to hydrogen would allow the storage of hydrogen at nearly 18 wt% compared to ~5% for gaseous hydrogen under pressure.

Rocket engines have also been fueled by ammonia. The Reaction Motors XLR99 rocket engine that powered the X-15 hypersonic research aircraft used liquid ammonia. Although not as powerful as other fuels, it left no soot in the reusable rocket engine and its density approximately matches the density of the oxidizer, liquid oxygen, which simplified the aircraft's design.

As a Stimulant

Anti-meth sign on tank of anhydrous ammonia, Otley, Iowa. Anhydrous ammonia is a common farm fertilizer that is also a critical ingredient in making methamphetamine. In 2005, Iowa state used grant money to give out thousands of locks to prevent criminals from getting into the tanks.

Ammonia, as the vapor released by smelling salts, has found significant use as a respiratory stimulant. Ammonia is commonly used in the illegal manufacture of methamphetamine through a Birch reduction. The Birch method of making methamphetamine is dangerous because the alkali metal and liquid ammonia are both extremely reactive, and the temperature of liquid ammonia makes it susceptible to explosive boiling when reactants are added.

Textile

Liquid ammonia is used for treatment of cotton materials, giving properties like mercerisation, using alkalis. In particular, it is used for prewashing of wool.

Lifting Gas

At standard temperature and pressure, ammonia is less dense than atmosphere, and has approximately 60% of the lifting power of hydrogen or helium. Ammonia has sometimes been used to fill weather balloons as a lifting gas. Because of its relatively high boiling point (compared to helium and hydrogen), ammonia could potentially be refrigerated and liquefied aboard an airship to reduce lift and add ballast (and returned to a gas to add lift and reduce ballast).

Woodworking

Ammonia has been used to darken quartersawn white oak in Arts & Crafts and Mission-style furniture. Ammonia fumes react with the natural tannins in the wood and cause it to change colours.

Safety Precautions

The world's longest ammonia pipeline, running from the TogliattiAzot
plant in Russia to Odessa in Ukraine.

The U. S. Occupational Safety and Health Administration (OSHA) has set a 15-minute exposure limit for gaseous ammonia of 35 ppm by volume in the environmental air and an 8-hour exposure limit of 25 ppm by volume. NIOSH recently reduced the IDLH from 500 to 300 based on recent more conservative interpretations of original research in 1943. IDLH (Immediately Dangerous to Life and Health) is the level to which a healthy worker can be exposed for 30 minutes without suffering irreversible health effects. Other organizations have varying exposure levels. U.S. Navy Standards [U.S. Bureau of Ships 1962] maximum allowable concentrations (MACs):continuous exposure (60 days): 25 ppm / 1 hour: 400 ppm Ammonia vapour has a sharp, irritating, pungent odour that acts as a warning of potentially dangerous exposure. The average odour threshold is 5 ppm, well below any danger or damage. Exposure to very high concentrations of gaseous ammonia can result in lung damage and death. Although ammonia is regulated in the United States as a non-flammable gas, it still meets the definition of a material that is toxic by inhalation and requires a hazardous safety permit when transported in quantities greater than 13,248 L (3,500 gallons). Household products containing ammonia (i.e., Windex) should never be used in conjunction with products containing bleach, as the resulting chemical reaction produces highly toxic fumes.

Liquid ammonia is dangerous because it is hygroscopic and because it can freeze flesh.

Toxicity

The toxicity of ammonia solutions does not usually cause problems for humans and other mammals, as a specific mechanism exists to prevent its build-up in the bloodstream. Ammonia is converted to carbamoyl phosphate by the enzyme carbamoyl phosphate synthetase, and then enters the urea cycle to be either incorporated into amino acids or excreted in the urine. Fish and amphibians lack this mechanism, as they can usually eliminate ammonia from their bodies by direct excretion. Ammonia even at dilute concentrations is highly toxic to aquatic animals, and for this reason it is classified as *dangerous for the environment*.

Coking Wastewater

Ammonia is present in coking wastewater streams, as a liquid by-product of the production of coke from coal. In some cases, the ammonia is discharged to the marine environment where it acts as a pollutant. The Whyalla steelworks in South Australia is one example of a coke-producing facility which discharges ammonia into marine waters.

Aquaculture

Ammonia toxicity is believed to be a cause of otherwise unexplained losses in fish hatcheries. Excess ammonia may accumulate and cause alteration of metabolism or increases in the body pH of the exposed organism. Tolerance varies among fish species. At lower concentrations, around 0.05 mg/L, un-ionised ammonia is harmful to fish species and can result in poor growth and feed conversion rates, reduced fecundity and fertility and increase stress and susceptibility to bacterial infections and diseases. Exposed to excess ammonia, fish may suffer loss of equilibrium, hyper-excitability, increased respiratory activity and oxygen uptake and increased heart rate. At concentrations exceeding 2.0 mg/L, ammonia causes gill and tissue damage, extreme lethargy, convulsions, coma, and death. Experiments have shown that the lethal concentration for a variety of fish species ranges from 0.2 to 2.0 mg/l.

During winter, when reduced feeds are administered to aquaculture stock, ammonia levels can be higher. Lower ambient temperatures reduce the rate of algal photosynthesis so less ammonia is removed by any algae present. Within an aquaculture environment, especially at large scale, there is no fast-acting remedy to elevated ammonia levels. Prevention rather than correction is recommended to reduce harm to farmed fish and in open water systems, the surrounding environment.

Storage Information

Similar to propane, anhydrous ammonia boils below room temperature when at atmospheric pressure. A storage vessel capable of 250 psi (1.7 MPa) is suitable to contain the liquid. Ammonium compounds should never be allowed to come in contact with bases (unless in an intended and contained reaction), as dangerous quantities of ammonia gas could be released.

Household Use

Solutions of ammonia (5–10% by weight) are used as household cleaners, particularly for glass. These solutions are irritating to the eyes and mucous membranes (respiratory and digestive tracts), and to a lesser extent the skin. Caution should be used that the chemical is never mixed into any liquid containing bleach, as a poisonous gas may result. Mixing with chlorine-containing products or strong oxidants, such as household bleach, can lead to hazardous compounds such as chloramines.

Laboratory use of Ammonia Solutions

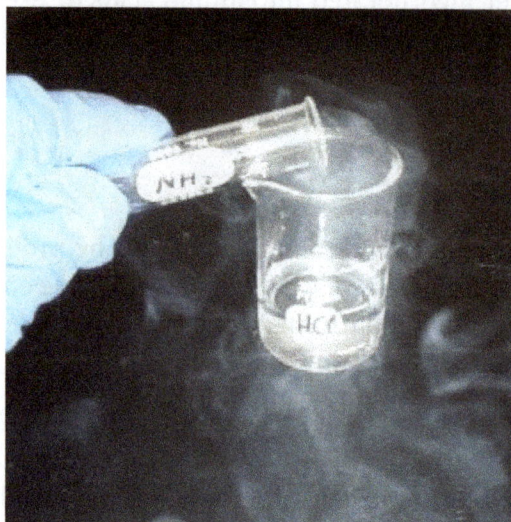

Hydrochloric acid sample releasing HCl fumes, which are reacting with ammonia fumes
to produce a white smoke of ammonium chloride.

The hazards of ammonia solutions depend on the concentration: "dilute" ammonia solutions are usually 5–10% by weight (<5.62 mol/L); "concentrated" solutions are usually prepared at >25% by weight. A 25% (by weight) solution has a density of 0.907 g/cm³, and a solution that has a lower density will be more concentrated. The European Union classification of ammonia solutions is given in the table.

Concentration by weight (w/w)	Molarity	Concentration mass/volume (w/v)	Classification	R-Phrases
5–10%	2.87–5.62 mol/L	48.9–95.7 g/L	Irritant (**Xi**)	R36/37/38
10–25%	5.62–13.29 mol/L	95.7–226.3 g/L	Corrosive (**C**)	R34
>25%	>13.29 mol/L	>226.3 g/L	Corrosive (**C**) Dangerous for the environment (**N**)	R34, R50

The ammonia vapour from concentrated ammonia solutions is severely irritating to the eyes and the respiratory tract, and these solutions should only be handled in a fume hood. Saturated ("0.880") solutions can develop a significant pressure inside a closed bottle in warm weather, and the bottle should be opened with care; this is not usually a problem for 25% ("0.900") solutions.

Ammonia solutions should not be mixed with halogens, as toxic and/or explosive products are formed. Prolonged contact of ammonia solutions with silver, mercury or iodide salts can also lead to explosive products: such mixtures are often formed in qualitative inorganic analysis, and should be lightly acidified but not concentrated (<6% w/v) before disposal once the test is completed.

Laboratory use of Anhydrous Ammonia (Gas or Liquid)

Anhydrous ammonia is classified as toxic (T) and dangerous for the environment (N). The gas is flammable (autoignition temperature: 651 °C) and can form explosive mixtures with air (16–25%).

The permissible exposure limit (PEL) in the United States is 50 ppm (35 mg/m³), while the IDLH concentration is estimated at 300 ppm. Repeated exposure to ammonia lowers the sensitivity to the smell of the gas: normally the odour is detectable at concentrations of less than 50 ppm, but desensitised individuals may not detect it even at concentrations of 100 ppm. Anhydrous ammonia corrodes copper- and zinc-containing alloys, and so brass fittings should not be used for handling the gas. Liquid ammonia can also attack rubber and certain plastics.

Ammonia reacts violently with the halogens. Nitrogen triiodide, a primary high explosive, is formed when ammonia comes in contact with iodine. Ammonia causes the explosive polymerisation of ethylene oxide. It also forms explosive fulminating compounds with compounds of gold, silver, mercury, germanium or tellurium, and with stibine. Violent reactions have also been reported with acetaldehyde, hypochlorite solutions, potassium ferricyanide and peroxides.

Synthesis and Production

Production trend of ammonia between 1947 and 2007

Because of its many uses, ammonia is one of the most highly produced inorganic chemicals. Dozens of chemical plants worldwide produce ammonia. Consuming more than 1% of all man-made power, ammonia production is a significant component of the world energy budget. The USGS reports global ammonia production in 2014 was 176 million tonnes. China accounted for 32.6% of that (increasingly from coal as part of urea synthesis), followed by Russia at 8.1%, India at 7.6%, and the United States at 6.4%. About 88% of the ammonia produced was used for fertilizing agricultural crops.

Before the start of World War I, most ammonia was obtained by the dry distillation of nitrogenous vegetable and animal waste products, including camel dung, where it was distilled by the reduction of nitrous acid and nitrites with hydrogen; in addition, it was produced by the distillation of coal, and also by the decomposition of ammonium salts by alkaline hydroxides such as quicklime, the salt most generally used being the chloride (sal ammoniac) thus:

$$2 NH_4Cl + 2 CaO \rightarrow CaCl_2 + Ca(OH)_2 + 2 NH_3$$

Hydrogen for ammonia synthesis could also be produced economically by using the water gas reaction followed by the water gas shift reaction, produced by passing steam through red-hot coke, to give a mixture of hydrogen and carbon dioxide gases, followed by removal of the carbon dioxide "washing" the gas mixture with water under pressure (25 standard atmospheres (2,500 kPa)); or by using other sources like coal or coke gasification.

Modern ammonia-producing plants depend on industrial hydrogen production to react with atmospheric nitrogen using a magnetite catalyst or over a promoted Fe catalyst under high pressure (100 standard atmospheres (10,000 kPa)) and temperature (450 °C) to form anhydrous liquid ammonia. This step is known as the ammonia synthesis loop (also referred to as the Haber–Bosch process):

$$3\,H_2 + N_2 \rightarrow 2\,NH_3$$

Hydrogen required for ammonia synthesis could also be produced economically using other sources like coal or coke gasification or less economically from the electrolysis of water into oxygen + hydrogen and other alternatives that are presently impractical for large scale. At one time, most of Europe's ammonia was produced from the Hydro plant at Vemork, via the electrolysis route. Various renewable energy electricity sources are also potentially applicable.

As a sustainable alternative to the relatively inefficient electrolysis, hydrogen can be generated from organic wastes (such as biomass or food-industry waste), using catalytic reforming. This releases hydrogen from carbonaceous substances at only 10–20% of energy used by electrolysis and may lead to hydrogen being produced from municipal wastes at below zero cost (allowing for the tipping fees and efficient catalytic reforming, such as cold-plasma). Catalytic (thermal) reforming is possible in small, distributed (even mobile) plants, to take advantage of low-value, stranded biomass/biowaste or natural gas deposits. Conversion of such wastes into ammonia solves the problem of hydrogen storage, as hydrogen can be released economically from ammonia on-demand, without the need for high-pressure or cryogenic storage.

It is also easier to store ammonia on board vehicles than to store hydrogen, as ammonia is less flammable than petrol or LPG.

For small scale laboratory synthesis, one can heat urea and $Ca(OH)_2$

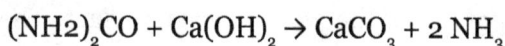

$$(NH2)_2CO + Ca(OH)_2 \rightarrow CaCO_3 + 2\,NH_3$$

Liquid Ammonia as a Solvent

Liquid ammonia is the best-known and most widely studied nonaqueous ionising solvent. Its most conspicuous property is its ability to dissolve alkali metals to form highly coloured, electrically conductive solutions containing solvated electrons. Apart from these remarkable solutions, much of the chemistry in liquid ammonia can be classified by analogy with related reactions in aqueous solutions. Comparison of the physical properties of NH_3 with those of water shows NH_3 has the lower melting point, boiling point, density, viscosity, dielectric constant and electrical conductivity; this is due at least in part to the weaker hydrogen bonding in NH_3 and because such bonding cannot form cross-linked networks, since each NH_3 molecule has only one lone pair of electrons compared with two for each H_2O molecule. The ionic self-dissociation constant of liquid NH_3 at −50 °C is about 10^{-33} mol^2·l^{-2}.

Solubility of salts

	Solubility (g of salt per 100 g liquid NH_3)
Ammonium acetate	253.2
Ammonium nitrate	389.6
Lithium nitrate	243.7
Sodium nitrate	97.6
Potassium nitrate	10.4
Sodium fluoride	0.35
Sodium chloride	157.0
Sodium bromide	138.0
Sodium iodide	161.9
Sodium thiocyanate	205.5

Liquid ammonia is an ionising solvent, although less so than water, and dissolves a range of ionic compounds, including many nitrates, nitrites, cyanides, thiocyanates, metal cyclopentadienyl complexes and metal bis(trimethylsilyl)amides. Most ammonium salts are soluble and act as acids in liquid ammonia solutions. The solubility of halide salts increases from fluoride to iodide. A saturated solution of ammonium nitrate contains 0.83 mol solute per mole of ammonia and has a vapour pressure of less than 1 bar even at 25 °C (77 °F).

Solutions of Metals

Liquid ammonia will dissolve the alkali metals and other electropositive metals such as magnesium, calcium, strontium, barium, europium and ytterbium. At low concentrations (<0.06 mol/l), deep blue solutions are formed: these contain metal cations and solvated electrons, free electrons that are surrounded by a cage of ammonia molecules.

These solutions are very useful as strong reducing agents. At higher concentrations, the solutions are metallic in appearance and in electrical conductivity. At low temperatures, the two types of solution can coexist as immiscible phases.

Redox Properties of Liquid Ammonia

	$E°$ (V, ammonia)	$E°$ (V, water)
$Li^+ + e^- \rightleftharpoons Li$	−2.24	−3.04
$K^+ + e^- \rightleftharpoons K$	−1.98	−2.93
$Na^+ + e^- \rightleftharpoons Na$	−1.85	−2.71
$Zn^{2+} + 2e^- \rightleftharpoons Zn$	−0.53	−0.76
$NH_4^+ + e^- \rightleftharpoons \frac{1}{2} H_2 + NH_3$	0.00	—
$Cu^{2+} + 2e^- \rightleftharpoons Cu$	+0.43	+0.34
$Ag^+ + e^- \rightleftharpoons Ag$	+0.83	+0.80

The range of thermodynamic stability of liquid ammonia solutions is very narrow, as the potential for oxidation to dinitrogen, $E°$ ($N_2 + 6NH_4^+ + 6e^- \rightleftharpoons 8NH_3$), is only +0.04 V. In practice, both oxidation to dinitrogen and reduction to dihydrogen are slow. This is particularly true of reducing solutions: the solutions of the alkali metals mentioned above are stable for several days, slowly decomposing to the metal amide and dihydrogen. Most studies involving liquid ammonia solutions are done in reducing conditions; although oxidation of liquid ammonia is usually slow, there is still a risk of explosion, particularly if transition metal ions are present as possible catalysts.

Ammonia's Role in Biological Systems and Human Disease

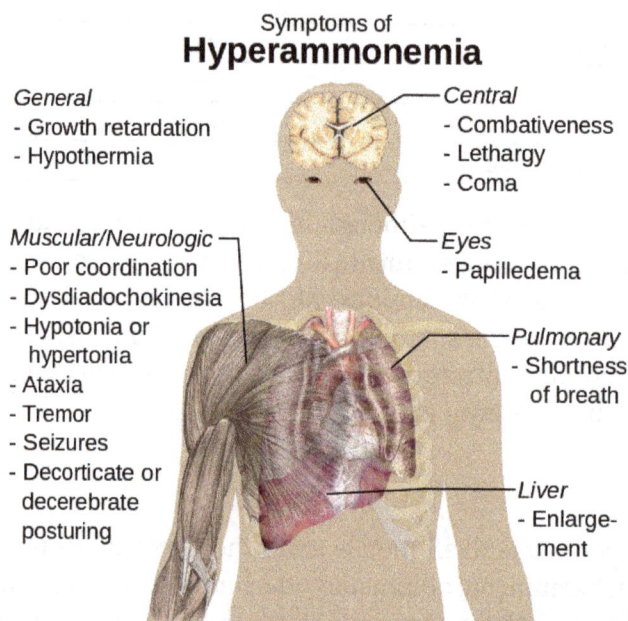

Symptoms of
Hyperammonemia

General
- Growth retardation
- Hypothermia

Central
- Combativeness
- Lethargy
- Coma

Muscular/Neurologic
- Poor coordination
- Dysdiadochokinesia
- Hypotonia or
  hypertonia
- Ataxia
- Tremor
- Seizures
- Decorticate or
  decerebrate
  posturing

Eyes
- Papilledema

Pulmonary
- Shortness
  of breath

Liver
- Enlarge-
  ment

Main symptoms of hyperammonemia (ammonia reaching toxic concentrations).

Ammonia is both a metabolic waste and a metabolic input throughout the biosphere. It is an important source of nitrogen for living systems. Although atmospheric nitrogen abounds (more than 75%), few living creatures are capable of using this atmospheric nitrogen in its diatomic form, N_2 gas. Therefore, nitrogen fixation is required for the synthesis of amino acids, which are the building blocks of protein. Some plants rely on ammonia and other nitrogenous wastes incorporated into the soil by decaying matter. Others, such as nitrogen-fixing legumes, benefit from symbiotic relationships with rhizobia that create ammonia from atmospheric nitrogen.

Biosynthesis

In certain organisms, ammonia is produced from atmospheric nitrogen by enzymes called nitrogenases. The overall process is called nitrogen fixation. Although it is unlikely that biomimetic methods that are competitive with the Haber process will be developed, intense effort has been directed toward understanding the mechanism of biological nitrogen fixation. The scientific interest in this problem is motivated by the unusual structure of the active site of the enzyme, which consists of an Fe_7MoS_9 ensemble.

Ammonia is also a metabolic product of amino acid deamination catalyzed by enzymes such as glutamate dehydrogenase 1. Ammonia excretion is common in aquatic animals. In humans, it is quickly converted to urea, which is much less toxic, particularly less basic. This urea is a major component of the dry weight of urine. Most reptiles, birds, insects, and snails excrete uric acid solely as nitrogenous waste.

In Physiology

Ammonia also plays a role in both normal and abnormal animal physiology. It is biosynthesised through normal amino acid metabolism and is toxic in high concentrations. The liver converts ammonia to urea through a series of reactions known as the urea cycle. Liver dysfunction, such as that seen in cirrhosis, may lead to elevated amounts of ammonia in the blood (hyperammonemia). Likewise, defects in the enzymes responsible for the urea cycle, such as ornithine transcarbamylase, lead to hyperammonemia. Hyperammonemia contributes to the confusion and coma of hepatic encephalopathy, as well as the neurologic disease common in people with urea cycle defects and organic acidurias.

Ammonia is important for normal animal acid/base balance. After formation of ammonium from glutamine, α-ketoglutarate may be degraded to produce two molecules of bicarbonate, which are then available as buffers for dietary acids. Ammonium is excreted in the urine, resulting in net acid loss. Ammonia may itself diffuse across the renal tubules, combine with a hydrogen ion, and thus allow for further acid excretion.

Excretion

Ammonium ions are a toxic waste product of metabolism in animals. In fish and aquatic invertebrates, it is excreted directly into the water. In mammals, sharks, and amphibians, it is converted in the urea cycle to urea, because it is less toxic and can be stored more efficiently. In birds, reptiles, and terrestrial snails, metabolic ammonium is converted into uric acid, which is solid, and can therefore be excreted with minimal water loss.

In Astronomy

Ammonia occurs in the atmospheres of the outer gas planets such as Jupiter (0.026% ammonia) and Saturn (0.012% ammonia).

Ammonia has been detected in the atmospheres of the gas giant planets, including Jupiter, along with other gases like methane, hydrogen, and helium. The interior of Saturn may include frozen crystals of ammonia. It is naturally found on Deimos and Phobos – the two moons of Mars.

Interstellar Space

Ammonia was first detected in interstellar space in 1968, based on microwave emissions from the direction of the galactic core. This was the first polyatomic molecule to be so detected. The sensitivity of the molecule to a broad range of excitations and the ease with which it can be observed in a number of regions has made ammonia one of the most important molecules for studies of molecular clouds. The relative intensity of the ammonia lines can be used to measure the temperature of the emitting medium.

The following isotopic species of ammonia have been detected:

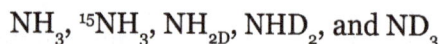

NH_3, $^{15}NH_3$, NH_{2D}, NHD_2, and ND_3

The detection of triply deuterated ammonia was considered a surprise as deuterium is relatively scarce. It is thought that the low-temperature conditions allow this molecule to survive and accumulate.

Since its interstellar discovery, NH_3 has proved to be an invaluable spectroscopic tool in the study of the interstellar medium. With a large number of transitions sensitive to a wide range of excitation conditions, NH_3 has been widely astronomically detected – its detection has been reported in hundreds of journal articles.

The study of interstellar ammonia has been important to a number of areas of research in the last few decades. Some of these are delineated below and primarily involve using ammonia as an interstellar thermometer.

Interstellar Formation Mechanisms

Ball-and-stick model of the diamminesilver(I) cation, $[Ag(NH_3)_2]^+$

The interstellar abundance for ammonia has been measured for a variety of environments. The $[NH_3]/[H_2]$ ratio has been estimated to range from 10^{-7} in small dark clouds up to 10^{-5} in the dense core of the Orion Molecular Cloud Complex. Although a total of 18 total production routes have been proposed, the principal formation mechanism for interstellar NH_3 is the reaction:

$NH_4^+ + e^- \rightarrow NH_3 + H\cdot$

The rate constant, k, of this reaction depends on the temperature of the environment, with a value of 5.2×10^{-6} at 10 K. The rate constant was calculated from the formula $k = a(T/300)^B$. For the primary formation reaction, $a = 1.05\times10^{-6}$ and $B = -0.47$. Assuming an NH_4^+ abundance of 3×10^{-7} and an electron abundance of 10^{-7} typical of molecular clouds, the formation will proceed at a rate of 1.6×10^{-9} cm^{-3}s^{-1} in a molecular cloud of total density 10^5 cm^{-3}.

All other proposed formation reactions have rate constants of between 2 and 13 orders of magnitude smaller, making their contribution to the abundance of ammonia relatively insignificant. As an example of the minor contribution other formation reactions play, the reaction:

$$H_2 + NH_2 \rightarrow NH_3 + H$$

has a rate constant of 2.2×10^{-15}. Assuming H_2 densities of 10^5 and $[NH_2]/[H_2]$ ratio of 10^{-7}, this reaction proceeds at a rate of 2.2×10^{-12}, more than 3 orders of magnitude slower than the primary reaction above.

Some of the other possible formation reactions are:

$$H^- + NH_4^+ \rightarrow NH_3 + H_2$$

$$PNH_3^+ + e^- \rightarrow P + NH_3$$

Interstellar Destruction Mechanisms

There are 113 total proposed reactions leading to the destruction of NH_3. Of these, 39 were tabulated in extensive tables of the chemistry among C, N, and O compounds. A review of interstellar ammonia cites the following reactions as the principal dissociation mechanisms:

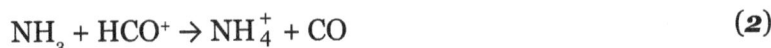

$$NH_3 + H_3^+ \rightarrow NH_4^+ + H_2 \qquad\qquad (1)$$

$$NH_3 + HCO^+ \rightarrow NH_4^+ + CO \qquad\qquad (2)$$

with rate constants of 4.39×10^{-9} and 2.2×10^{-9}, respectively. The above equations (1, 2) run at a rate of 8.8×10^{-9} and 4.4×10^{-13}, respectively. These calculations assumed the given rate constants and abundances of $[NH_3]/[H_2] = 10^{-5}$, $[H_3^+]/[H_2] = 2\times10^{-5}$, $[HCO^+]/[H_2] = 2\times10^{-9}$, and total densities of $n = 10^5$, typical of cold, dense, molecular clouds. Clearly, between these two primary reactions, equation (1) is the dominant destruction reaction, with a rate ~10,000 times faster than equation (2). This is due to the relatively high abundance of H_3^+.

Single Antenna Detections

Radio observations of NH_3 from the Effelsberg 100-m Radio Telescope reveal that the ammonia line is separated into two components – a background ridge and an unresolved core. The background corresponds well with the locations previously detected CO. The 25 m Chilbolton telescope in England detected radio signatures of ammonia in H II regions, HNH_2O masers, H-H objects, and other objects associated with star formation. A comparison of emission line widths indicates that turbulent or systematic velocities do not increase in the central cores of molecular clouds.

Microwave radiation from ammonia was observed in several galactic objects including $W_3(OH)$, Orion A, W43, W51, and five sources in the galactic centre. The high detection rate indicates that this is a common molecule in the interstellar medium and that high-density regions are common in the galaxy.

Interferometric Studies

VLA observations of NH_3 in seven regions with high-velocity gaseous outflows revealed condensations of less than 0.1 pc in L1551, S140, and Cepheus A. Three individual condensations were detected in Cepheus A, one of them with a highly elongated shape. They may play an important role in creating the bipolar outflow in the region.

Extragalactic ammonia was imaged using the VLA in IC 342. The hot gas has temperatures above 70 K, which was inferred from ammonia line ratios and appears to be closely associated with the innermost portions of the nuclear bar seen in CO. NH_3 was also monitored by VLA toward a sample of four galactic ultracompact HII regions: G9.62+0.19, G10.47+0.03, G29.96-0.02, and G31.41+0.31. Based upon temperature and density diagnostics, it is concluded that in general such clumps are likely to be the sites of massive star formation in an early evolutionary phase prior to the development of an ultracompact HII region.

Infrared Detections

Absorption at 2.97 micrometres due to solid ammonia was recorded from interstellar grains in the Becklin-Neugebauer Object and probably in NGC 2264-IR as well. This detection helped explain the physical shape of previously poorly understood and related ice absorption lines.

A spectrum of the disk of Jupiter was obtained from the Kuiper Airborne Observatory, covering the 100 to 300 cm^{-1} spectral range. Analysis of the spectrum provides information on global mean properties of ammonia gas and an ammonia ice haze.

A total of 149 dark cloud positions were surveyed for evidence of 'dense cores' by using the (J,K) = (1,1) rotating inversion line of NH_3. In general, the cores are not spherically shaped, with aspect ratios ranging from 1.1 to 4.4. It is also found that cores with stars have broader lines than cores without stars.

Ammonia has been detected in the Draco Nebula and in one or possibly two molecular clouds, which are associated with the high-latitude galactic infrared cirrus. The finding is significant because they may represent the birthplaces for the Population I metallicity B-type stars in the galactic halo that could have been borne in the galactic disk.

Observations of Nearby Dark Clouds

By balancing and stimulated emission with spontaneous emission, it is possible to construct a relation between excitation temperature and density. Moreover, since the transitional levels of ammonia can be approximated by a 2-level system at low temperatures, this calculation is fairly simple. This premise can be applied to dark clouds, regions suspected of having extremely low temperatures and possible sites for future star formation. Detections of ammonia in dark clouds show very narrow lines—indicative not only of low temperatures, but also of a low level of inner-cloud turbulence. Line ratio calculations provide a measurement of cloud temperature that is independent of previous CO observations. The ammonia observations were consistent with CO measurements of rotation temperatures of ~10 K. With this, densities can be determined, and have been calculated to range between 10^4 and 10^5 cm^{-3} in dark clouds. Mapping of NH_3 gives typical clouds sizes of 0.1 pc and masses near 1 solar mass. These cold, dense cores are the sites of future star formation.

UC HII Regions

Ultra-compact HII regions are among the best tracers of high-mass star formation. The dense material surrounding UCHII regions is likely primarily molecular. Since a complete study of massive star formation necessarily involves the cloud from which the star formed, ammonia is an invaluable tool in understanding this surrounding molecular material. Since this molecular material can be spatially resolved, it is possible to constrain the heating/ionising sources, temperatures, masses, and sizes of the regions. Doppler-shifted velocity components allow for the separation of distinct regions of molecular gas that can trace outflows and hot cores originating from forming stars.

Extragalactic Detection

Ammonia has been detected in external galaxies, and by simultaneously measuring several lines, it is possible to directly measure the gas temperature in these galaxies. Line ratios imply that gas temperatures are warm (~50 K), originating from dense clouds with sizes of tens of pc. This picture is consistent with the picture within our Milky Way galaxy—hot dense molecular cores form around newly forming stars embedded in larger clouds of molecular material on the scale of several hundred pc (giant molecular clouds; GMCs).

Ammonia, as used commercially, is often called anhydrous ammonia. This term emphasizes the absence of water in the material. Because NH_3 boils at −33.34 °C (−28.012 °F) at a pressure of 1 atmosphere, the liquid must be stored under high pressure or at low temperature. "Household ammonia" or "ammonium hydroxide" is a solution of NH_3 in water. The concentration of such solutions is measured in units of the Baumé scale (density), with 26 degrees baumé (about 30% w/w ammonia at 15.5 °C) being the typical high-concentration commercial product. Household ammonia ranges in concentration from 5 to 10 weight percent ammonia.

Approximately 83% (as of 2004) of ammonia is used as fertilizers either as its salts or as solutions. When applied to soil, it helps provide increased yields of crops such as corn and wheat. Consuming more than 1% of all man-made power, the production of ammonia is a significant component of the world energy budget.

Physical Properties of Ammonia

Ammonia has flammable limits of 16%-25% by volume in the air and 15%-79% in oxygen. Ammonia –air mixtures is 650 Celsius. The mixture can explode if ignited. Ammonia is readily soluble in water. A large amount of heat, about 2180kj (520kcal), is produced when the dissolution of 1kg of ammonia gas occurs.

Physical Properties of Anhydrous Ammonia

Property	Unit	Value
Molecular weight	1	17.03
Boiling point	°C	-33.35
Freezing point	°C	-77.7
Critical temperature	°C	133.0
Critical pressure	MPa	11.425
Specific heat (gas)	j/kg °K	
-at 0°C		2,097.2
-at 100 °C		2,226.2
-at 200 °C		2,105.6
Heat of formation (gas)	kj/mol	
-at 0 °K		-39.2
-at 298 °K		-46.2
Solubility in water	%, weight	
-at 0 °C		42.8
-at 20 °C		33.1
-at 40 °C		23.4
-at 60 °C		4.1
Specific gravity	g/ml	
-at -40 °C		0.690
-at 0 °C		0.639
-at 40 °C		0.580

Density of Ammonia at 15°C

Ammonia (wt %)	Density (g/cm³)
8	0.970
16	0.947
32	0.889
50	0.832
75	0.733
100	0.618

Production of Ammonia

Ammonia is produced in a process known as the Haber process, in which nitrogen and hydrogen react in the presence of an iron catalyst to form ammonia. The hydrogen is formed by reacting natural gas and steam at high temperatures and the nitrogen is supplied from the air. Other gases (such as water and carbon dioxide) are removed from the gas stream and the nitrogen and hydrogen passed over an iron catalyst at high temperature and pressure to form the ammonia. The process is shown schematically in the figure below.

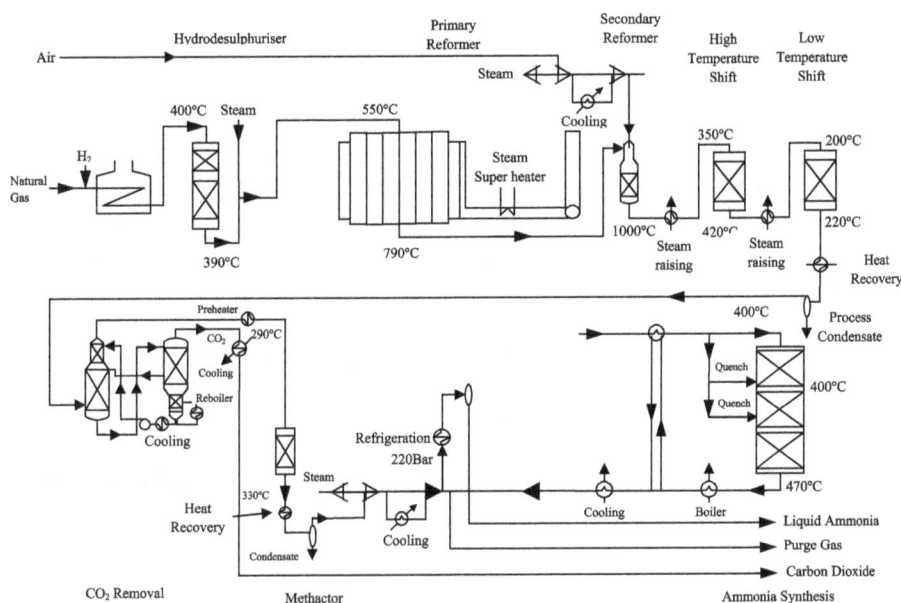

Industrial manufacturing of Ammonia

Step 1 - Hydrogen Production

Hydrogen is produced by the reaction of methane with water. However, before this can be carried out, all sulfurous compounds must be removed from the natural gas to prevent catalyst poisoning. These are removed by heating the gas to 400°C and reacting it with zinc oxide:

$$ZnO + H_2S \rightarrow ZnS + H_2O$$

Following this, the gas is sent to the primary reformer for steam reforming, where superheated steam is fed into the reformer with the methane. The gas mixture heated with natural gas and purge gas to 770°C in the presence of a nickel catalyst. At this temperature the following equilibrium reactions are driven to the right, converting the methane to hydrogen, carbon dioxide and small quantities of carbon monoxide:

$$CH_4 + H_2O \rightarrow 3H_2 + CO$$

$$CH_4 + 2H_2O \rightarrow 4H_2 + CO_2$$

$$CO + H_2O \rightarrow H_2 + CO_2$$

This gaseous mixture is known as synthesis gas.

Step 2 - Nitrogen Addition

The synthesis gas is cooled slightly to 735°C. It then flows to the secondary reformer whereit is mixed with a calculated amount of air. The highly exothermic reaction between oxygen and methane produces more hydrogen. Important reactions are:

$$CO + H_2O \rightarrow CO_2 + H_2$$

$$O_2 + 2CH_4 \rightarrow 2CO + 4H_2$$

$$O_2 + CH_4 \rightarrow CO_2 + 2H_2$$

$$2O_2 + CH_4 \rightarrow 2H_2O + CO_2$$

In addition, the necessary nitrogen is added in the secondary reformer. As the catalyst that is used to form the ammonia is pure iron, water, carbon dioxide and carbon monoxide must be removed from the gas stream to prevent oxidation of the iron. This is carried out in the next three steps.

The Ammonia production process is continued here from step 3 onwards.

Step 3 - Removal of Carbon Monoxide

Here the carbon monoxide is converted to carbon dioxide (which is used later in the synthesis of urea) in a reaction known as the water gas shift reaction:

$$CO + H_2O \rightarrow CO_2 + H_2$$

This is achieved in two steps. Firstly, the gas stream is passed over a Cr/Fe_3O_4 catalyst at 360°C and then over a Cu/ZnO/Cr catalyst at 210°C. The same reaction occurs in both steps, but using the two steps maximises conversion.

Step 4 - Water Removal

The gas mixture is further cooled to 40°C, at which temperature the water condenses out and is removed.

Step 5 - Removal of Carbon Oxides

The gases are then pumped up through a counter-current of UCARSOL D solution (an MDEA solution). Carbon dioxide is highly soluble in UCARSOL, and more than 99.9% of the CO_2 in the mixture dissolves in it. The remaining CO_2 (as well as any CO that was not converted to CO_2 in Step 3) is converted to methane (methanation) using

a Ni/Al_2O_3 catalyst at 325°C: $2CO + 3H_2 \rightarrow CH_4 + H_2O$

$$CO_2 + 4H_2 \rightarrow CH_4 + 2H_2O$$

The water which is produced in these reactions is removed by condensation at 40°C as above. The carbon dioxide is stripped from the UCARSOL and used in urea manufacture. The UCARSOL is cooled and reused for carbon dioxide removal.

Step 6 - Synthesis of Ammonia

The gas mixture is now cooled, compressed and fed into the ammonia synthesis loop. A mixture of ammonia and unreacted gases which have already been around the loop are mixed with the incoming gas stream and cooled to 5°C. The ammonia present is removed and the unreacted gases heated to 400°C at a pressure of 330 barg and passed over an iron catalyst. Under these conditions 26% of the hydrogen and nitrogen are converted to ammonia. The outlet gas from the ammonia converter is cooled from 220°C to 30°C. This cooling process condenses more the half the ammonia, which is then separated out. The remaining gas is mixed with more cooled, compressed incoming gas. The reaction occuring in the ammonia converter is:

$$N_2 + 3H_2 \rightarrow 2NH_3$$

The ammonia is rapidly decompressed to 24 barg. At this pressure, impurities such as methane and hydrogen become gases. The gas mixture above the liquid ammonia (which also contains significant levels of ammonia) is removed and sent to the ammonia recovery unit. This is an absorber-stripper system using water as solvent. The remaining gas (purge gas) is used as fuel for the heating of the primary reformer. The pure ammonia remaining is mixed with the pure ammonia from the initial condensation above and is ready for use in urea production, for storage or for direct sale.

Description of Storage and Transfer Equipment

Liquefied ammonia from production plants is either used directly in downstream plants or transferred to storage tanks. From these the ammonia can be transferred to road tankers, rail tank cars or ships.

Ammonia is usually stored by using one or other of three methods:-

- Fully refrigerated storage in large tanks with a typical capacity of 10,000 to 30,000 tonnes (up to 50,000)

- Pressurised storage spheres or cylinders up to about 1,700 tonnes

- Semi-refrigerated tanks

Emissions during normal operation are negligible. Major leaks of ammonia from storage tanks are almost unknown with most of the leaks which do occur being during transport or transfer.

A well designed, constructed, operated and maintained installation has a very low probability of an ammonia leak of hazardous proportions. However, even though the residual risk is small, the effects of a major leak on areas of high population density could be very serious. It is therefore good practice to build ammonia storage and handling installations at a sufficient distance from domestic housing, schools, hospitals or any area where substantial numbers of people may assemble. In some countries there are planning procedures or regulations which control the siting of ammonia storage installations and similar establishments. Where there are no formal controls, the siting of ammonia storage facilities should be given serious consideration at the design stage. It is undesirable for ammonia storage tanks to be sited close to installations where there is a risk of fire or explosion, since these could increase the possibility of a release of ammonia.

Storage Tanks

Anhydrous ammonia is stored in three types of tank as outlined above:-

- Fully refrigerated at a temperature of about −33°C, these tanks are provided with refrigeration equipment

- Non-refrigerated tanks in which the ammonia is stored at ambient emperature

- Semi-refrigerated spheres

Refrigerated storage is preferred for storage of large quantities of liquid ammonia. The initial release of ammonia in the case of a line or tank failure is much slower than with pressurized ammonia.

There are several construction types for the storage of refrigerated liquid products. The most important types are:-

- Single containment: a single-wall insulated tank, normally with a containment bund around it

- Double containment: this type of storage tank has two vertical walls, both of which are designed to contain the stored amount of liquid and withstand the hydrostatic pressure of the liquid. The roof rests on the inner wall

- Full containment: the two walls of this closed storage tank are also designed to contain the stored amount of liquid, but in this case the roof rests on the outer wall. The tank must be constructed in conformity with an agreed code for the construction of pressure vessels or storage tanks and taking account of its pressure and operating temperature.

The design and materials of construction of the tank should be checked by consulting an appropriate national, or recognised international, standard. These could make demands on the blast resistance of storage tanks in some cases. The storage tank must be safeguarded against high pressure and in the case of refrigerated liquid ammonia also against a pressure below the minimum design pressure. The ingress of warm ammonia into cold ammonia must be avoided to eliminate risk of excessive evaporation and the "roll-over" phenomenon. All storage tanks should be equipped with two independent level indicators, each having a high level alarm.

An automatic cut-off valve, operated by a very high level alarm instrument, should be installed on the feeding line. In cases of refrigerated liquid ammonia, storage tanks must be equipped with a recompression installation to liquefy the boil-off. There should be at least two refrigeration units to allow proper maintenance and to prevent the emission of ammonia via the relief valves.

Furthermore, an installed alternative power supply may be necessary. An automatic discharge system to a flare may be provided in case of failure of the refrigeration equipment. The flare must be located at a suitable distance from the tanks. Relief valves should be provided, appropriate for the duty using an adequate margin between operating and relief pressure.

Nitric Acid

Nitric acid (HNO_3), also known as aqua fortis and spirit of niter, is a highly corrosive mineral acid.

The pure compound is colorless, but older samples tend to acquire a yellow cast due to decomposition into oxides of nitrogen and water. Most commercially available nitric acid has a concentration of 68% in water. When the solution contains more than 86% HNO_3, it is referred to as *fuming nitric acid*. Depending on the amount of nitrogen dioxide present, fuming nitric acid is further characterized as white fuming nitric acid or red fuming nitric acid, at concentrations above 95%.

Nitric acid is the primary reagent used for nitration – the addition of a nitro group, typically to an organic molecule. While some resulting nitro compounds are shock- and thermally-sensitive explosives, a few are stable enough to be used in munitions and demolition, while others are still more stable and used as pigments in inks and dyes. Nitric acid is also commonly used as a strong oxidizing agent.

Physical and Chemical Properties

Commercially available nitric acid is an azeotrope with water at a concentration of 68% HNO_3, which is the ordinary concentrated nitric acid of commerce. This solution has a boiling temperature of 120.5 °C at 1 atm. Two solid hydrates are known; the monohydrate ($HNO_3 \cdot H_2O$) and the trihydrate ($HNO_3 \cdot 3H_2O$).

Nitric acid 70%

Nitric acid of commercial interest usually consists of the maximum boiling azeotrope of nitric acid and water, which is approximately 68% HNO_3, (approx. 15 molar). This is considered concentrated or technical grade, while reagent grades are specified at 70% HNO_3. The density of concentrated nitric acid is 1.42 g/cm³. An older density scale is occasionally seen, with concentrated nitric acid specified as 42° Baumé.

Contamination with Nitrogen Dioxide

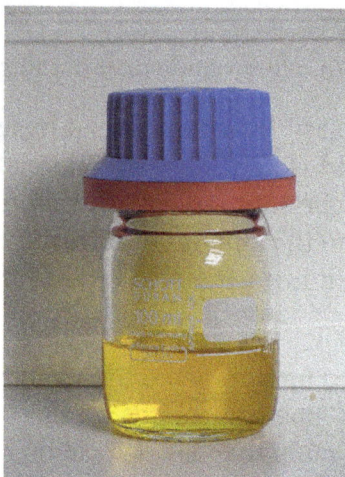

Fuming nitric acid contaminated with yellow nitrogen dioxide.

Nitric acid is subject to thermal or light decomposition and for this reason it was often stored in brown glass bottles: $4\,HNO_3 \rightarrow 2\,H_2O + 4\,NO_2 + O_2$. This reaction may give rise to some non-negligible variations in the vapor pressure above the liquid because the nitrogen oxides produced dissolve partly or completely in the acid.

The nitrogen dioxide (NO_2) remains dissolved in the nitric acid coloring it yellow or even red at higher temperatures. While the pure acid tends to give off white fumes when exposed to air, acid with dissolved nitrogen dioxide gives off reddish-brown vapors, leading to the common name "red fuming acid" or "fuming nitric acid" – the most concentrated form of nitric acid at Standard Temperature and Pressure (STP). Nitrogen oxides (NO_x) are soluble in nitric acid.

Fuming Nitric Acid

A commercial grade of fuming nitric acid contains 90% HNO_3 and has a density of 1.50 g/cm³. This grade is often used in the explosives industry. It is not as volatile nor as corrosive as the anhydrous acid and has the approximate concentration of 21.4 molar.

Red fuming nitric acid, or RFNA, contains substantial quantities of dissolved nitrogen dioxide (NO_2) leaving the solution with a reddish-brown color. Due to the dissolved nitrogen dioxide, the density of red fuming nitric acid is lower at 1.490 g/cm³.

An *inhibited* fuming nitric acid (either IWFNA, or IRFNA) can be made by the addition of 0.6 to 0.7% hydrogen fluoride (HF). This fluoride is added for corrosion resistance in metal tanks. The fluoride creates a metal fluoride layer that protects the metal.

Anhydrous Nitric Acid

White fuming nitric acid, pure nitric acid or WFNA, is very close to anhydrous nitric acid. It is available as 99.9% nitric acid by assay. One specification for white fuming nitric acid is that it has a maximum of 2% water and a maximum of 0.5% dissolved NO_2. Anhydrous nitric acid has a density of 1.513 g/cm³ and has the approximate concentration of 24 molar. Anhydrous nitric acid is a colorless mobile liquid with a density of 1.512 g/cm³, which solidifies at −42 °C to form white crystals. As it decomposes to NO_2 and water, it obtains a yellow tint. It boils at 83 °C. It is usually stored in a glass shatterproof amber bottle with twice the volume of head space to allow for pressure build up. When received, the pressure must be released and repeated monthly until finished.

Structure and Bonding

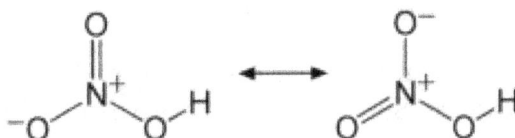

Two major resonance representations of HNO_3

The molecule is planar. Two of the N–O bonds are equivalent and relatively short (this can be explained by theories of resonance; the canonical forms show double-bond character in these two bonds, causing them to be shorter than typical N–O bonds), and the third N–O bond is elongated because the O atom is also attached to a proton.

Reactions

Acid-base Properties

Nitric acid is normally considered to be a strong acid at ambient temperatures. There is some disagreement over the value of the acid dissociation constant, though the pK_a value is usually reported as less than −1. This means that the nitric acid in diluted solution is fully dissociated except in extremely acidic solutions. The pK_a value rises to 1 at a temperature of 250 °C.

Nitric acid can act as a base with respect to an acid such as sulfuric acid:

$$HNO_3 + 2\,H_2SO_4 \rightleftharpoons NO_2^+ + H_3O^+ + 2IISO_4^-; \text{ Equillibrium constant: } K \sim 22$$

The nitronium ion, NO_2^+, is the active reagent in aromatic nitration reactions. Since nitric acid has both acidic and basic properties, it can undergo an autoprotolysis reaction, similar to the self-ionization of water:

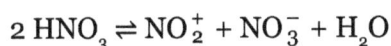

$$2\,HNO_3 \rightleftharpoons NO_2^+ + NO_3^- + H_2O$$

Reactions with Metals

Nitric acid reacts with most metals, but the details depend on the concentration of the acid and the nature of the metal. Dilute nitric acid behaves as a typical acid in its reaction with most metals. Magnesium, manganese and zinc liberate H_2:

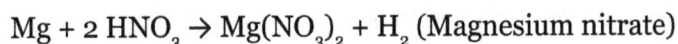

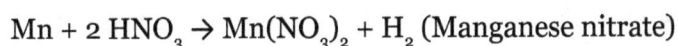

$$Mg + 2\,HNO_3 \rightarrow Mg(NO_3)_2 + H_2 \text{ (Magnesium nitrate)}$$

$$Mn + 2\,HNO_3 \rightarrow Mn(NO_3)_2 + H_2 \text{ (Manganese nitrate)}$$

$$Zn + 2\,HNO_3 \rightarrow Zn(NO_3)_2 + H_2 \text{ (Zinc nitrate)}$$

Nitric acid can oxidize non-active metals such as copper and silver. With these non-active or less electropositive metals the products depend on temperature and the acid concentration. For example, copper reacts with dilute nitric acid at ambient temperatures with a 3:8 stoichiometry:

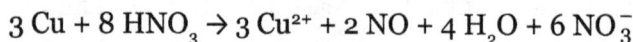

$$3\,Cu + 8\,HNO_3 \rightarrow 3\,Cu^{2+} + 2\,NO + 4\,H_2O + 6\,NO_3^-$$

The nitric oxide produced may react with atmospheric oxygen to give nitrogen dioxide. With more concentrated nitric acid, nitrogen dioxide is produced directly in a reaction with 1:4 stoichiometry:

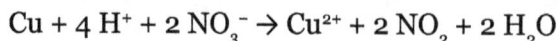

$$Cu + 4\,H^+ + 2\,NO_3^- \rightarrow Cu^{2+} + 2\,NO_2 + 2\,H_2O$$

Upon reaction with nitric acid, most metals give the corresponding nitrates. Some metalloids and metals give the oxides; for instance, Sn, As, Sb, and Ti are oxidized into SnO_2, As_2O_5, Sb_2O_5, and TiO_2 respectively.

Some precious metals, such as pure gold and platinum-group metals do not react with nitric acid, though pure gold does react with *aqua regia*, a mixture of concentrated nitric acid and hydrochloric acid. However, some less noble metals (Ag, Cu, ...) present in some gold alloys relatively poor in gold such as colored gold can be easily oxidized and dissolved by nitric acid, leading to colour changes of the gold-alloy surface. Nitric acid is used as a cheap means in jewelry shops to quickly spot low-gold alloys (< 14 carats) and to rapidly assess the gold purity.

Being a powerful oxidizing agent, nitric acid reacts violently with many non-metallic compounds, and the reactions may be explosive. Depending on the acid concentration, temperature and the reducing agent involved, the end products can be variable. Reaction takes place with all metals except the noble metals series and certain alloys. As a general rule, oxidizing reactions occur primarily with the concentrated acid, favoring the formation of nitrogen dioxide (NO_2). However, the powerful oxidizing properties of nitric acid are thermodynamic in nature, but sometimes its oxidation reactions are rather kinetically non-favored. The presence of small amounts of nitrous acid (HNO_2) greatly enhance the rate of reaction.

Although chromium (Cr), iron (Fe), and aluminium (Al) readily dissolve in dilute nitric acid, the concentrated acid forms a metal-oxide layer that protects the bulk of the metal from further oxidation. The formation of this protective layer is called passivation. Typical passivation concentrations range from 20% to 50% by volume. Metals that are passivated by concentrated nitric acid are iron, cobalt, chromium, nickel, and aluminium.

Reactions with Non-metals

Being a powerful oxidizing acid, nitric acid reacts violently with many organic materials and the reactions may be explosive. The hydroxyl group will typically strip a hydrogen from the organic molecule to form water, and the remaining nitro group takes the hydrogen's place. Nitration of organic compounds with nitric acid is the primary method of synthesis of many common explosives, such as nitroglycerin and trinitrotoluene (TNT). As very many less stable byproducts are possible, these reactions must be carefully thermally controlled, and the byproducts removed to isolate the desired product.

Reaction with non-metallic elements, with the exceptions of nitrogen, oxygen, noble gases, silicon, and halogens other than iodine, usually oxidizes them to their highest oxidation states as acids with the formation of nitrogen dioxide for concentrated acid and nitric oxide for dilute acid.

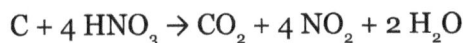

$$C + 4 HNO_3 \rightarrow CO_2 + 4 NO_2 + 2 H_2O$$

or

$$3 C + 4 HNO_3 \rightarrow 3 CO_2 + 4 NO + 2 H_2O$$

Concentrated nitric acid oxidizes I_2, P_4, and S_8 into HIO_3, H_3PO_4, and H_2SO_4, respectively.

Xanthoproteic Test

Nitric acid reacts with proteins to form yellow nitrated products. This reaction is known as the xanthoproteic reaction. This test is carried out by adding concentrated nitric acid to the substance being tested, and then heating the mixture. If proteins that contain amino acids with aromatic rings are present, the mixture turns yellow. Upon adding a base such as ammonia, the color turns orange. These color changes are caused by nitrated aromatic rings in the protein. Xanthoproteic acid is formed when the acid contacts epithelial cells. Respective local skin color changes are indicative of inadequate safety precautions when handling nitric acid.

Production

Nitric acid is made by reaction of nitrogen dioxide (NO_2) with water.

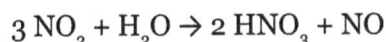

$$3 NO_2 + H_2O \rightarrow 2 HNO_3 + NO$$

Normally, the nitric oxide produced by the reaction is reoxidized by the oxygen in air to produce additional nitrogen dioxide.

Bubbling nitrogen dioxide through hydrogen peroxide can help to improve acid yield.

$$2 NO_2 + H_2O_2 \rightarrow 2 HNO_3$$

Commercial grade nitric acid solutions are usually between 52% and 68% nitric acid. Production of nitric acid is via the Ostwald process, named after German chemist Wilhelm Ostwald. In this process, anhydrous ammonia is oxidized to nitric oxide, in the presence of platinum or rhodium gauze catalyst at a high temperature of about 500 K and a pressure of 9 bar.

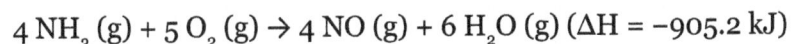

$$4 NH_3 (g) + 5 O_2 (g) \rightarrow 4 NO (g) + 6 H_2O (g) (\Delta H = -905.2 \text{ kJ})$$

Nitric oxide is then reacted with oxygen in air to form nitrogen dioxide.

$$2 NO (g) + O_2 (g) \rightarrow 2 NO_2 (g) (\Delta H = -114 \text{ kJ/mol})$$

This is subsequently absorbed in water to form nitric acid and nitric oxide.

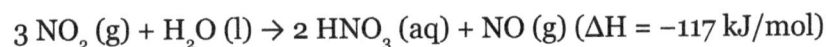

$$3 NO_2 (g) + H_2O (l) \rightarrow 2 HNO_3 (aq) + NO (g) (\Delta H = -117 \text{ kJ/mol})$$

The nitric oxide is cycled back for reoxidation. Alternatively, if the last step is carried out in air:

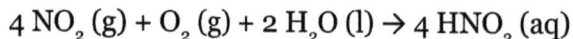

$$4 \, NO_2 \, (g) + O_2 \, (g) + 2 \, H_2O \, (l) \rightarrow 4 \, HNO_3 \, (aq)$$

The aqueous HNO_3 obtained can be concentrated by distillation up to about 68% by mass. Further concentration to 98% can be achieved by dehydration with concentrated H_2SO_4. By using ammonia derived from the Haber process, the final product can be produced from nitrogen, hydrogen, and oxygen which are derived from air and natural gas as the sole feedstocks.

Prior to the introduction of the Haber process for the production of ammonia in 1913, nitric acid was produced using the Birkeland–Eyde process, also known as the arc process. This process is based upon the oxidation of atmospheric nitrogen by atmospheric oxygen to nitric oxide at very high temperatures. An electric arc was used to provide the high temperatures, and yields of up to 4% nitric oxide were obtained. The nitric oxide was cooled and oxidized by the remaining atmospheric oxygen to nitrogen dioxide, and this was subsequently absorbed in dilute nitric acid. The process was very energy intensive and was rapidly displaced by the Ostwald process once cheap ammonia became available.

Laboratory Synthesis

In laboratory, nitric acid can be made by thermal decomposition of copper(II) nitrate, producing nitrogen dioxide and oxygen gases, which are then passed through water to give nitric acid.

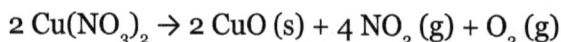

$$2 \, Cu(NO_3)_2 \rightarrow 2 \, CuO \, (s) + 4 \, NO_2 \, (g) + O_2 \, (g)$$

An alternate route is by reaction of approximately equal masses of any nitrate salt such as sodium nitrate with 96% sulfuric acid (H_2SO_4), and distilling this mixture at nitric acid's boiling point of 83 °C. A nonvolatile residue of the metal sulfate remains in the distillation vessel. The red fuming nitric acid obtained may be converted to the white nitric acid.

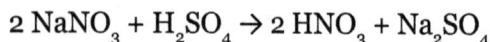

$$2 \, NaNO_3 + H_2SO_4 \rightarrow 2 \, HNO_3 + Na_2SO_4$$

The dissolved NO_x are readily removed using reduced pressure at room temperature (10–30 min at 200 mmHg or 27 kPa) to give white fuming nitric acid. This procedure can also be performed under reduced pressure and temperature in one step in order to produce less nitrogen dioxide gas.

Dilute nitric acid may be concentrated by distillation up to 68% acid, which is a maximum boiling azeotrope containing 32% water. In the laboratory, further concentration involves distillation with either sulfuric acid or magnesium nitrate which act as dehydrating agents. Such distillations must be done with all-glass apparatus at reduced pressure, to prevent decomposition of the acid. Industrially, highly concentrated nitric acid is produced by dissolving additional nitrogen dioxide in 68% nitric acid in an absorption tower. Dissolved nitrogen oxides are either stripped in the case of white fuming nitric acid, or remain in solution to form red fuming nitric acid. More recently, electrochemical means have been developed to produce anhydrous acid from concentrated nitric acid feedstock.

Uses

The main industrial use of nitric acid is for the production of fertilizers. Nitric acid is neutralized with ammonia to give ammonium nitrate. This application consumes 75–80% of the 26M tons produced annually (1987). The other main applications are for the production of explosives, nylon precursors, and specialty organic compounds.

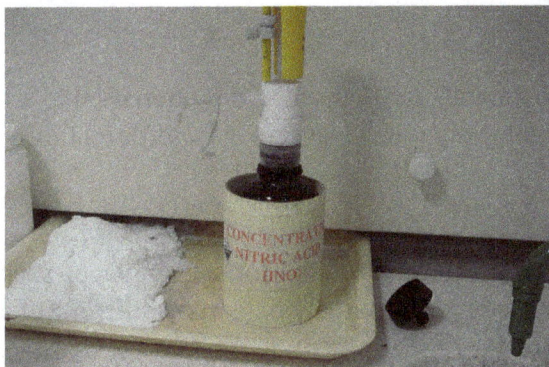

Nitric acid in a laboratory

Precursor to Organic Nitrogen Compounds

In organic synthesis, industrial and otherwise, the nitro group is a versatile functional group. Most derivatives of aniline are prepared via nitration of aromatic compounds followed by reduction. Nitrations entail combining nitric and sulfuric acids to generate the nitronium ion, which electrophilically reacts with aromatic compounds such as benzene. Many explosives, such as TNT, are prepared this way.

Use as an Oxidant

The precursor to nylon, adipic acid, is produced on a large scale by oxidation of cyclohexanone and cyclohexanol with nitric acid.

Rocket Propellant

Nitric acid has been used in various forms as the oxidizer in liquid-fueled rockets. These forms include red fuming nitric acid, white fuming nitric acid, mixtures with sulfuric acid, and these forms with HF inhibitor. IRFNA (inhibited red fuming nitric acid) was one of 3 liquid fuel components for the BOMARC missile.

Niche uses

Analytical Reagent

In elemental analysis by ICP-MS, ICP-AES, GFAA, and Flame AA, dilute nitric acid (0.5 to 5.0%) is used as a matrix compound for determining metal traces in solutions. Ultrapure trace metal grade acid is required for such determination, because small amounts of metal ions could affect the result of the analysis.

It is also typically used in the digestion process of turbid water samples, sludge samples, solid samples as well as other types of unique samples which require elemental analysis via ICP-MS, ICP-OES, ICP-AES, GFAA and flame atomic absorption spectroscopy. Typically these digestions use a 50% solution of the purchased HNO_3 mixed with Type 1 DI Water.

In electrochemistry, nitric acid is used as a chemical doping agent for organic semiconductors, and in purification processes for raw carbon nanotubes.

Woodworking

In a low concentration (approximately 10%), nitric acid is often used to artificially age pine and maple. The color produced is a grey-gold very much like very old wax or oil finished wood (wood finishing).

Etchant and Cleaning Agent

The corrosive effects of nitric acid are exploited for a number of specialty applications, such as etching in printmaking, pickling stainless steel or cleaning silicon wafers in electronics.

A solution of nitric acid, water and alcohol, Nital, is used for etching of metals to reveal the microstructure. ISO 14104 is one of the standards detailing this well known procedure.

Commercially available aqueous blends of 5–30% nitric acid and 15–40% phosphoric acid are commonly used for cleaning food and dairy equipment primarily to remove precipitated calcium and magnesium compounds (either deposited from the process stream or resulting from the use of hard water during production and cleaning). The phosphoric acid content helps to passivate ferrous alloys against corrosion by the dilute nitric acid.

Nitric acid can be used as a spot test for alkaloids like LSD, giving a variety of colours depending on the alkaloid.

Safety

Second degree burn caused by nitric acid Nitric acid is a corrosive acid and a powerful oxidizing agent. The major hazard posed by it is chemical burns as it carries out acid hydrolysis with proteins (amide) and fats (ester) which consequently decomposes living tissue (e.g. skin and flesh). Concentrated nitric acid stains human skin yellow due to its reaction with the keratin. These yellow stains turn orange when neutralized. Systemic effects are unlikely, however, and the substance is not considered a carcinogen or mutagen.

The standard first aid treatment for acid spills on the skin is, as for other corrosive agents, irrigation with large quantities of water. Washing is continued for at least ten to fifteen minutes to cool the tissue surrounding the acid burn and to prevent secondary damage. Contaminated clothing is removed immediately and the underlying skin washed thoroughly.

Being a strong oxidizing agent, reactions of nitric acid with compounds such as cyanides, carbides, or metallic powders can be explosive and those with many organic compounds, such as turpentine, are violent and hypergolic (i.e. self-igniting). Hence, it should be stored away from bases and organics.

History

The first mention of nitric acid is in Pseudo-Geber's *De Inventione Veritatis*, wherein it is obtained by calcining a mixture of niter, alum and blue vitriol. It was again described by Albert the Great in the 13th century and by Ramon Lull, who prepared it by heating niter and clay and called it "eau forte" (aqua fortis).

Glauber devised a process to obtain it by distillate potassium nitrate with sulfuric acid. In 1776 Lavoisier showed that it contained oxygen, and in 1785 Henry Cavendish determined its precise composition and showed that it could be synthesized by passing a stream of electric sparks through moist air.

Preparation of Nitric Acid

All plants for the production of nitric acid are currently based on the same basic chemical operations: Oxidation of ammonia with air to give nitric oxide and oxidation of the nitric oxide to nitrogen dioxide and absorption in water to give a solution of nitric acid. The efficiency of the first step is favoured by low pressure whereas that of the second is favoured by high pressure. These considerations, combined with economic reasons give rise to two types of nitric acid plant, single pressure plants and dual pressure plants. In the single pressure plant, the oxidation and absorption steps take place at essentially the same pressure. In dual pressure plants absorption takes place at a higher pressure than the oxidation stage.

The oxidation and absorption steps can be classified as: low pressure (pressure below 1,7 bar), medium pressure (1,7 - 6,5 bar) and high pressure (6,5 - 13 bar). Except for some very old plants, single pressure plants operate at medium or high pressure and dual pressure plants operate at medium pressure for the oxidation stage and high pressure for the absorption.

The main unit operation involved in the nitric acid process are the same for all types of plant and in sequential order these are: air filtration, air compression, air/ammonia mixing, air/ammonia oxidation over catalytic gauzes, energy recovery by steam generation and/or gas re-heating, gas cooling, gas compression, energy recovery and cooling (dual pressure plants only), absorption with the production of nitric acid, waste gas (tail gas) heating and energy recovery by expansion of the waste gas to atmosphere in a gas turbine.

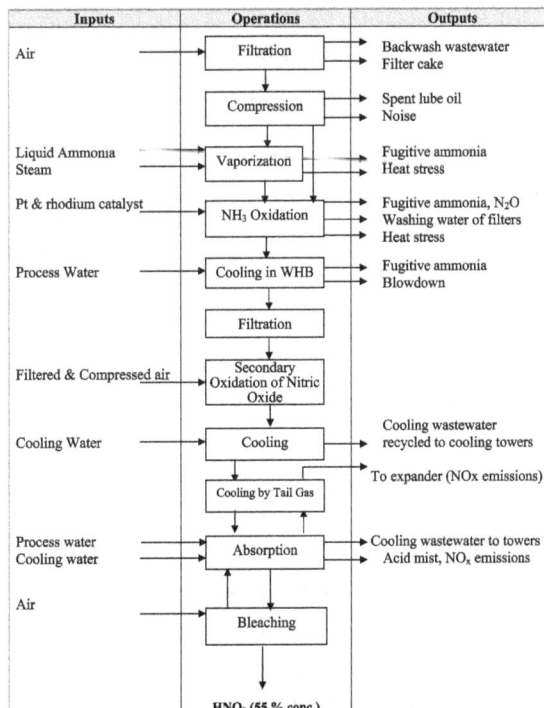

Process Flow Diagram for Nitric Acid Manufacturing

Although there are three main processes in the nitric acid production (mono, dual and atmospheric pressure) but the routes are more or less the same as follows:

1. Primary Air Filtration

Feed air contains, beside nitrogen and oxygen, some inert gases, carbon monoxide and some dust and impurities. Dust and impurities harmfully affect the platinum catalyst efficiency, therefore air must be filtered through a series of filters. Air is sucked (by Nox turbo- compressor set in case of atmospheric or by an air compressor in case of mono or dual pressure processes) through a filter (usually candle felt filter elements in case of atmospheric or special paper filter elements) and then heated.

2. Air Preheating:

In this step filtered air is introduced through a series of steel pipes, in which fledges are provided on its outer surface to increase the heating surface area. Steam, at 3 kgm/cm2, 140°C is passed through the pipes and air outside. The steam condensate is collected by steam traps and recycled to the boilers.

3. Ammonia Evaporation and Filtration

Anhydrous liquid ammonia is evaporated, superheated and its pressure is maintained according to the process (whether atmospheric or pressure). The ammonia is filtered (usually across ceramic candle filters that are cleaned every now and then).

4. Mixture Filtration

The purpose of this operation is to increase purification of (air/ammonia) mixture. Cylindrical filters of large diameters are used, each one contains a porous ceramic candle hanged on a disc with openings. These candles permit the mixture to pass leaving the impurities on their surfaces. After a certain level of impurities, the filters needed to be opened and washed with water then dried by hot air. The generated wastewater from washing the filters generate a pollution problem.

5. Ammonia Oxidation (converters)

Ammonia is reacted with air on platinum/ rhodium alloy catalysts in the oxidation section of nitric acid plants. Nitric oxide and water are formed in this process. The yield of nitric oxide depend on pressure and temperature:

- Pressure below 1,7 bar, temperature 810-850°C -> NO yield 97%

- Pressure 1,7-6,5 bar, temperature 850-900°C -> NO yield 96%

- Pressure above 6,5 bar, temperature 900-940°C -> NO yield 95%

In this operation combustion takes place with the aid of platinum rhodium catalyst. It consists of several woven or knitted gauzes formed from wire containing about 90 % platinum alloyed with rhodium for greater strength and sometimes containing palladium. Heated ammonia (60° C) is mixed with air (at 80° C) in a pipe with a big diameter. The ammonia/ air mixture is introduced to the converter. The converter contains two parts, the upper part (of a conical shape) has a pierced disc fitted on its entrance to distribute the mixture on the platinum net equally. The lower part of the converter is a bottom pierced pot filled with pottery rings to distribute the hot gases (evolving from the reaction) on to the coils of the bottom boiler. The ammonia air ratio should be strictly maintained at 12 % in case of atmospheric process or 13 % in case of dual pressure process.

Ammonia goes through oxidation according to the following reaction

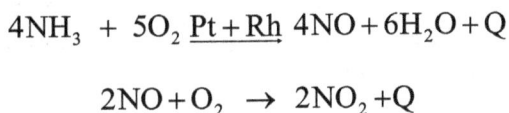

$$4NH_3 \ + \ 5O_2 \ \underline{Pt+Rh} \ 4NO+6H_2O+Q$$

$$2NO+O_2 \ \rightarrow \ 2NO_2+Q$$

Temperature is adjusted in a range 800°-900°C, because above 900 nitrogen oxide will decompose to N_2 and O_2, and below 800° C nitrogen oxide will be formed which does not produce nitric acid when dissolved in water. The heat released from those highly exothermic reactions is mostly recovered by the waste heat boiler (WHB) fitted in the ammonia reaction (burners) in the form of superheated steam for running the NOx and air compressors.

Air pollution and contamination from the ammonia can poison the catalyst. This effect, as well as poor ammonia-air mixing and poor gas distribution across the catalyst, may reduce the yield by 10 %. Maintenance of the catalyst operating temperature is very important for the NO yield. This is achieved by adjusting the air/ammonia ratio and ensuring that the lower explosive limit for ammonia in air is not exceeded. The preheated ammonia is thoroughly mixed with preheated air and subjected to further filtration to avoid contaminants from entering to the catalyst. The ratio of ammonia air mixture is controlled by a high precision ratio controller which is considered as the safeguard for protecting arising from:

1. Catalyst temperature exceeds 850° C

2. Ammonia/ air ratio exceeds 12.5 % (in atmospheric oxidation) or 10 % (in case of pressure oxidation)

3. Failure of air compressor or Nox compressor ...etc

The water produced in oxidation is then condensed in a cooler-condenser and transferred to the absorption column. Nitric oxide is oxidized to nitrogen dioxide as the combustion gases are cooled. For this purpose secondary air is added to the gas mixture obtained from the ammonia oxidation to increase the oxygen content to such a level that the waste gas leaving the plant has a normal oxygen content of between 2 - 4 %.

6. Energy Recovery

The hot reaction gases are used to produce steam and/or to preheat the waste gas (tail gas). The heated waste gas is discharged to the atmosphere through a gas turbine for energy recovery. The combustion gas after this heat transfer for energy recovery, has a temperature of 100 to 200°C, depending on the process and it is then further cooled with water.

i. Cooling

Exit gases (NOx, excess air and water vapour) are cooled through a water cooler in two stages, where nitrogen oxides are then dissolved in the water condensate to form a very diluted nitric acid (about 2% conc.) which is collected in a tank. The collected diluted acid is withdrawn by pumps and discharged to the bottom of the absorption tower. The remaining gases are withdrawn, together with excess air, by compressing turbines to the absorption towers.

ii. Absorption:

The absorber is operated with a counter-current flow of water. the absorption of the nitrogen dioxide and its reaction to nitric acid and nitric oxide take place simultaneously in the gaseous and liquid phases. The main reaction taking place is as follows:

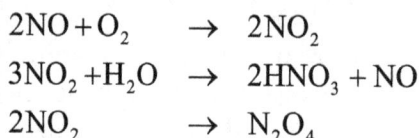

$$2NO + O_2 \rightarrow 2NO_2$$
$$3NO_2 + H_2O \rightarrow 2HNO_3 + NO$$
$$2NO_2 \rightarrow N_2O_4$$

These reactions depend on pressure and temperature to a large extent and are favored by higher pressure and lower temperature.

Series of towers with 3.5m diameter and 24m height are used in this stage. The absorber- reactor is a sieve plate or bubble cap unit with cooling coils on each of the 20 to 50 trays. Gas enters at the bottom, dilute nitric acid part way up the column, with cold water entering at the top. In the first tower most of the oxidation operation takes place where NO is converted to nitrogen dioxide. Diluted Nitric acid is then formed, collected in a level tank and discharged to the bleaching operation. Chlorine impurities presents a unique problem in the absorber, they cannot be transferred through the bottom neither can they leave in the top gas. Therefore, they must be excluded from entry or provision made for their purging as the reaction of nitric oxide proceeds during gas flow. Consequently the gases are cooled (remove heat of reaction and promote oxidation reactions to go to completion), against the cold tail gases coming out of the absorption tower to preheat them before entering the tail gas turbine. The preheated tail gas passes through the expansion turbine (part of the NOx compressor drives) to recover the energy and reduce tail gas temperature. Then further cooled before passing through the series of counter-current packed towers.

iii. Bleaching

The acid leaving the bottom of the column contains some NOx, mostly as N_2O_4 (colorless) but some as red NO_2. It is therefore sent to a bleaching tower to eliminate its colour. In bleaching operation the acid is sprayed at the top of the tower (about 5.5m height and 90 cm diameter), which is filled with a Raschig rings at the bottom. A counter current air stream is introduced at the bottom of the tower through a screen, to absorb the gases from the acid then withdrawn from the top of the tower. The acid, free from gases, is collected at the bottom of the tower and then cooled in cooling plates and sent to storage tanks. Cooling plates are essential, especially at the first tower, where the highest quantity of heat is released because most of the oxidation and absorption reactions take place in that tower.

Major Hazards

The following hazards may arise during nitric acid production:

- Equipment/piping failure because of corrosion

- Explosion hazard due to the air ammonia mixture

- Explosion of nitrite/nitrate salts.

Corrosion protection is achieved by the well proven use of suitable austenitic stainless steel where condensation can occur and by regular monitoring of conditions. Safety is ensured by the automatic closure of the ammonia control valve and separate shutdown trip valve when too high an air ammonia ratio is measured, either from each individual flow meter or indirectly from the catalyst gauze temperature. The air ammonia ratio should be continuously controlled and kept below the hazardous range. Any free ammonia present in the nitrous gas will have a deposit of nitrite/nitrate in a cold spot. Local washing and well proven operating practices will prevent the hazard.

Platinum Recovery

During operation the surface of the catalyst is damaged by abrasion and vaporization. Vaporization loss dominates at the beginning of operation, but vaporization weakness the metal structure and leads to abrasion and erosion.

Platinum from the catalyst passes into the gas stream in the form of very fine particles, and its loss can substantially increase the production cost. Therefore, several methods of platinum recovery were develop and installed in many plants. Two types of recovery systems – catchment gauzes and mechanical filters – are usually offered.

The principle of catchment gauzes is to collect platinum at a temperature as high as possible while the main portion of the platinum loss is still in vapor form. At these temperatures, platinum atoms strike the metal surface and form an alloy with the catchment metal for subsequent recovery. The system can recover up to 80% of the platinum losses. The catchment gauzes, which are installed at the bottom of the burner, are composed of a mesh screen and two or more metal gauzes. Catchment gauzes are returned together with the catalyst gauzes to the precious metal refining plant.

The mechanical filters, which are composed of glass wool or silica fibers, are commonly installed downstream of the catalyst where the gas temperature is below 300°C. Recovery rates of 50% have been reported.

Construction Materials

The corrosive behavior of nitric acid toward metals requires the proper selection of construction material. The principal material wherever nitric acid or wet nitric oxides are present is chromium-nickel austenitic steel. The carbon content in this steel must be kept as low as possible because chromium forms carbides that are not acid resistant. Alloyed steels are also used for welded parts of pumps, impellers, and rotating elements of compressors. For equipment that handles ammonia, air, and hot, dry gases, normal carbon steel can be used. However, for safety in operation, especially during startup and shutdown operations, nitric acid plants are often equipped with practically all stainless steel equipment. Because they are resistant to nitric acid, various fluorocarbon plastic materials are used for flanges, gaskets, and seals.

Industrial Processes

Industrial plants are classified according to the pressure used.

- Monopressure plants. These plants operate at the same pressure in the ammonia oxidation and absorption sections of the plant. Monopressure plants are classified as low pressure (0.3-0.5 MPa) and high pressure (0.8-1.3 MPa).

- Dual-pressure plants. These plants operate under lower pressure in the ammonia oxidation section than in the absorption section.

In general high-pressure operation permits smaller plant units to be used for a given output and helps to reduce capital costs. High pressure also favor NO_2 absorption; however, high- pressure ammonia oxidation induces greater catalyst losses and also increases power requirements unless additional equipment is installed for power recovery. Because of recent emphasis on pollution control, the ability of high-pressure processes to attain acceptably low Nox levels in the tail-gas has favored their adoption. Therefore, most new plants use either mono high-pressure or dual-pressure (medium-pressure combustion, High-pressure absorption) process although some mono-medium-pressure processes are used.

The first nitric acid plants used near-atmospheric pressure throughout. However, practically all modern plants use elevated pressures. The choice of the process always depends on local conditions; however, a general principle may be that lower capacity plants favor the high monopressure alternative. The dual-pressure choice seems to be a compromise between the higher investment costs of mono-medium pressure and higher operating costs of mono-high pressure alternative.

Ammonium Nitrate

Ammonium Nitrate is in the first place a nitrogenous fertilizer representing more than 10% of the total nitrogen consumption worldwide. It is more readily available to crops than urea. In the second place, due to its powerful oxidizing properties is used with proper additives as commercial explosive.

It is applied as a straight material or in combination with calcium carbonate, limestone, or dolomite. The combination is called calcium ammonium nitrate (CAN) or ammonium nitrate-limestone (ANL) or various trade names and in compound fertilizers including nitrophosphates. It is also a principal ingredient of most liquid nitrogen fertilizers. The nitrogen in ammonium nitrate is more rapidly available to some crops than urea or ammonium sulfate; most crops take up nitrogen mainly in nitrate form; thus, ammoniacal nitrogen must be converted to nitrate in the soil before it becomes effective. Many crops respond well to a mixture of ammonium and nitrate nitrogen. Even though the nitrification process is rapid in warm soil, it is slower in cool soil (10°C and below). Urea may cause seedling damage due to volatilization of ammonia, and ammonium sulfate is strongly acid forming. For these and reasons ammonium nitrate and CAN are effective fertilizers in zones with medium and low temperatures during the cropping period, especially in those with a short vegetation period.

The main advantages of ammonium nitrate are as follows:

- It is quite hygroscopic;

- There is some risk of fire or even explosions unless suitable precautions are taken;

- It is less effective for flooded rice than urea or ammoniacal nitrozen fertilizers;

- It is more prone to leaching immediately after application than ammoniacal products.

Ammonium Nitrate Properties

Ammonium nitrate is a white crystalline substance with a nitrogen content of 35% and a density of about 1.725 kg/m³. The melting points depend on the content of the water, and it is practically impossible to obtain dry product in the industrial conditions.

In production and storage of ammonium nitrate, transformations of the crystal states that may affect the quality of the product occur.

Some additives such as $Mg(NO_3)_2$, (NH_4) $2SO_4$, and some others can slightly change the critical relative humidities.

Ammonium Nitrate Production

The production process comprises three main unit operations: neutralization, evaporation, solidification (prilling and granulation). Individual plants vary widely in process detail.

Neutralization

Anhydrous liquid ammonia is evaporated in an evaporator using cooling water. The stoichiometic quantities of nitric acid (55% concentration wt/ wt) and gaseous ammonia are introduced by an automatic ratio controller to a neutralizer. The reaction between Ammonia and nitric acid produces ammonium nitrate solution according to the following exothermic reaction.

$$NH_3 + HNO_3 \rightarrow NH_4 NO_3$$

Neutralization can be performed in a single stage or in two stages. The neutralizer can be carried out at atmospheric (either normal or low emission neutralizers where the temperature does not exceed 105°C and pH will be 6 and 3 respectively) or at elevated pressure of almost 4 atmospheres. The normal neutralizers are usually followed by flash evaporation in order to in crease the out let A.N concentration to 70%. In case of pressure neutralizers the temperature will be in the range of 178°C and the steam generated from the heat of reaction will be utilized in the subsequent step namely concentration of A.N solution.

During evaporation some ammonia is lost from the solution. The steam which is boiled off is contaminated. The control of the neutralizer is important. The pH and the temperature must both be strictly controlled to limit the losses from the neutralizer. All installations must include pH and temperature controls. At the operating temperature of the neutralizer, impurity control is of great importance because a safety incident will also be a significant environmental incident. The ammonium nitrate solution from neutralizer may be fed to storage without further processing but, if it is used in the manufacture of solid ammonium nitrate, it is concentrated by evaporation.

Evaporation to Concentrate the A.N

The outlet from the neutralizer is received in an intermediate tank. The solution should be made alkaline before being pumped (no need for pumps in case of pressure neutralizers since the pressure will maintain the flow) to the evaporation section (multi-effect) running under vacuum. The solution will be steam heated in the multi effect evaporation section. The solution will be concentrated up to 97.5-99.5% (normally over 99 %) depending on whether ammonium nitrate will be granulated or prilled.

Mixing the Filling Material

In order to reduce the nitrogen content of A.N from 35% to 33.5%, the proper filling material is added (about 4% by weight of powdered limestone or dolomite or even kaolin).

Prilling or Granulation

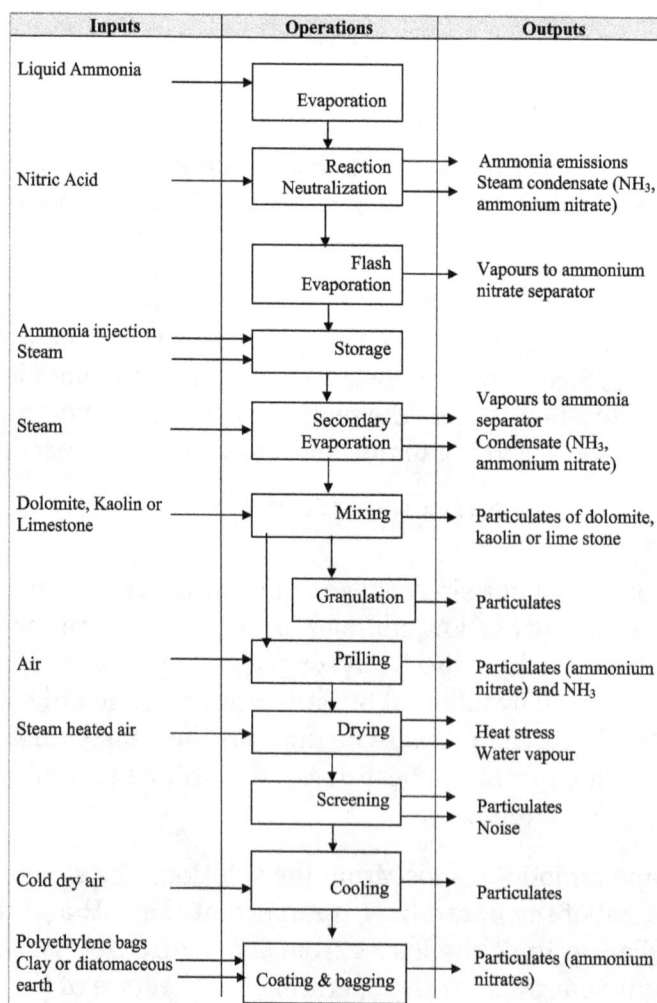

Inputs	Operations	Outputs
Liquid Ammonia	Evaporation	
Nitric Acid	Reaction Neutralization	Ammonia emissions / Steam condensate (NH_3, ammonium nitrate)
	Flash Evaporation	Vapours to ammonium nitrate separator
Ammonia injection / Steam	Storage	
Steam	Secondary Evaporation	Vapours to ammonia separator / Condensate (NH_3, ammonium nitrate)
Dolomite, Kaolin or Limestone	Mixing	Particulates of dolomite, kaolin or lime stone
	Granulation	Particulates
Air	Prilling	Particulates (ammonium nitrate) and NH_3
Steam heated air	Drying	Heat stress / Water vapour
	Screening	Particulates / Noise
Cold dry air	Cooling	Particulates
Polyethylene bags / Clay or diatomaceous earth	Coating & bagging	Particulates (ammonium nitrates)

Process Flow Diagram for Ammonium Nitrate Manufacturing

The hot concentrated melt is either granulated (fluidize bed granulation, drum granulation, etc) or prilled. Ammonium nitrate is formed into droplets which then fall down a fall tower (prill tower) where they cool and solidify. Granulation requires more complicated plant than prilling and variety of equipment. The main advantage of granulation with respect of environment is that the quantity of air to be treated is much smaller and abatement equipment is cheaper.

Drying, Screening

The ammonium nitrate (prills or granules) is dried (usually in drums) using hot air (steam heated), then screened to separate the correct product size. The oversize and undersize will be recycled either in the mixing tank (in case of prilling) or to the granulator.

Final Cooling

The hot proper size granules, are then cooled (against cooled and humid free air) down to 40°C and treated with anti-caking (usually amines) and then coated with an inert material (usually, kaolin, limestone or dolomite) and then conveyed to the storage.

Major Hazards

Ammonia, nitric acid and ammonium nitrate are the hazardous chemicals present in ammonium nitrate plants. A.N is an oxidizing agent and precautions must be taken in manufacturing, transport and storage.

The main chemical hazards associated with ammonium nitrate are fire, decomposition and explosion. Burns caused by hot AN solution should also be considered from a safety point of view.

Ammonium nitrate itself does not burn. Being an oxidizing agent, it can facilitate the initiation of a fire and intensify fires in combustible materials. Hot AN solution can initiate a fire in rags, wooden articles etc., on coming into contact with them. Similarly, fertilizer products or dust contaminated with oil or other combustible materials can also start fires when left on hot surfaces.

Pure solid A.N melts at 169°C. On further heating it decomposes by way of a complex series of reactions. Up to about 250°C it decomposes primarily into N_2O and H_2O. Above 300°C reactions producing N_2, NO, NO_2 etc., become significant. These reactions are exothermic and irreversible. They are accompanied by the vapour pressure dependent endothermic dissociation into HNO_3 and NH_3 vapours which can provide a temperature limiting mechanism, provided the gases can escape freely. If they cannot, the endothermic dissociation is suppressed and a run-away decomposition can develop, leading to explosive behavior.

A number of materials have a strong catalytic effect on the thermal decomposition of A.N. These include acids, chlorides, organic materials, chromates, dichromate, salts of manganese, copper and nickel and certain metals such as zinc, copper and lead. The decomposition of AN is suppressed or prevented by an alkaline condition. Thus the addition of ammonia offers a major safeguard against the decomposition hazard. The release of toxic fumes is one of the main hazards associated with the decomposition of AN.

Strongly acidic conditions and the presence of contaminants should be avoided to counter the explosion hazard in AN solutions. Explosions can occur when ammonium nitrate is heated under confinement in pumps. Reasons for pump explosions include:

1. No (or insufficient) flow through the pump.

2. incorrect design (design may incorporate low flow and/or high temperature trips).

3. poor maintenance practices.

4. contamination.

Product Handling

Ammonium nitrate may be stored in bulk although in most climates this requires air-conditioned facilities. Storage facilities should have adequate ventilation to allow quick dispersion of heat and toxic gases in the event of fire The storage area must be equipped with a high efficiency sprinkler system. Ammonium nitrate has its own oxygen for burning, and only large amounts of water can extinguish a fire. In most countries the commercial product is bagged in bags that should be "moisture proof"; at least one ply should be impermeable to moisture. If they are properly designed, plastic-film bags or bags with plastic liners are suitable. Bulk shipment is common in some countries using covered, hopper-bottom cars. Before loading, the inside of the vehicle should be thoroughly cleaned. It is important to prevent contamination of ammonium nitrate with organic materials such as grease or other hydrocarbons, chlorates, nitrates, and metal salts (Zn, Cu), which when ignited may support fire.

Production of Straight Granulated AN and CAN

Products with nitrogen contents in the range of 30%-34.5% are classified as straight ammonium nitrate. The following granulation processes are offered for straight ammonium nitrate production:

- Cold spherodizers

- Drum granulation

- Pugmill

- Pan granulation

- Fluid bed granulation

- Fluidized drum granulation

These processes are used for products containing up to 34.5%N. For products that contain up to 33.5% N, the pugmill, drum and cold spherodizers are used also [30,31]. These processes operate the same way as for CAN. However, the ammonium nitrate melt concentrations (wt. %) are different:

- Drum   96-96.5

- Pugmill   95.5-96

- Pan granulation   99.5

- Fluid bed granulation   99

- Fluidized drum granulation   97.5

With all methods, additives are obligatory for granulation and for improving storage properties.

Dry recycled material is fed at a controlled rate to the inclined rotating pan granulator. In the granulator a hot melt, which is virtually moisture free, is sprayed onto the moving bed of solids and solidifies on the cool particles. Round granules are formed by agglomeration, and, as their size increases, they move upwards in the rotating pan, finally rolling over the rim. The granulation temperature is controlled by the rate at which solids are fed to the pan. An optimum temperature range for agglomeration , within which a high growth rate of the particles is obtained, is 5°-25°C below the temperature at which the fertilizer melt solidifies. The recycle ratio under these conditions is about 0.5/0.7:1 for ammonium nitrate. Granules leaving the pan are plastic and have a somewhat irregular surface. They enter a polishing drum where they are exposed to mild mechanical forces and smoothed. A certain amount of cooling also occurs. Cooling to the desired product temperature is then performed in normal cooling equipment, such as fluidized bed or a rotary drum. Depending on the climatic conditions at the plant site and the desired product temperature, the cooling air may be conditioned. Cooled granules are conveyed to a screen. Oversize material from the screen is fed to a crusher and the crushed material, undersize granules, and dust from the cyclones, are recycled to the pan.

Because of the high melt concentration and temperature, a "blue fume" of submicron ammonium nitrate- similar to the fume during prilling of high-density ammonium nitrate – has to be recovered for treatment. However, the air flow from the pan and polishing drum is relatively small. A wet scrubber may be used for recovery; this is not usually practical in a prilling process with a high rate of air flow from the top of the prilling tower. The air from the product cooler is treated in wet or dry cyclones.

Pollution Control

As in all industrial operations, pollution control requirements for AN and CAN plants have become more stringent in recent years. This has posed a difficult problem for high- density AN prilling because of the large volume of air exhausted from prill towers and because of the very small particle size of the fume in the air. Fuming is much more severe in high-density prilling because the AN melt must be at higher temperature (about 180°C) to keep it from freezing. At this temperature there is an appreciable vapor pressure of $NH_3 + HNO_3$ resulting from dissociation of AN, according to the equation:

$$NH_4NO_3 \rightarrow NH_3 + HNO_3$$

The dissociation products recombine in the cooler air to form a blue haze consisting of AN particles of submicron size. Particles of this size are difficult to collect, and they present a highly visible and stable haze or "blue fume." The problem is much less serious with low-density prilling because of lower AN solution temperatures. It is less serious in granulation processes because of much smaller volumes of air in contact with hot solution.

Production of Calcium Ammonium Nitrate

Calcium ammonium nitrate, often abbreviated CAN, is a fertilizer which is a blend of about 20%-30% $CaCO_3$ and 70%-80% Ammonium nitrate. CAN is produced by mixing concentrated ammonium nitrate solution with ground calcitic or dolomitic limestone, chalk marl, or precipitated calcium carbonate from nitrophosphate production. The mixing should be done quickly to avoid decomposition of the ammonium nitrate:

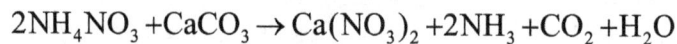

$$2NH_4NO_3 + CaCO_3 \rightarrow Ca(NO_3)_2 + 2NH_3 + CO_2 + H_2O$$

Both technologies – prilling and granulation – can be used for production of CAN. In prilling CAN the AN solution is premixed with the ground limestone immediately before prilling. A rotating perforated bucket is the preferred type of drop-forming instrument. Prill towers are very high (30 m-50 m) depending on the AN solution concentration and cooling equipment used. Prilled CAN is conditioned with china clay, kieselguhur or calcined Fuller's earth in amount ranging from 1% to 3%. In the prill tower mean particle size is 2 mm-2.5 mm. To obtain larger product, Hoechst and CDF Chimie (now AZF) have developed combined prilling-granulation technologies.

The prilling tower produces only seed prills from about 35% of the ammonium nitrate solution. The prills are directed to the swelling drum where the rest of the AN and calcium carbonate are added to the AN solution. The fines are recycled.

The following granulation processes are available:

Cold spherodizer., Fluid bed., Pugmill.and Drum.

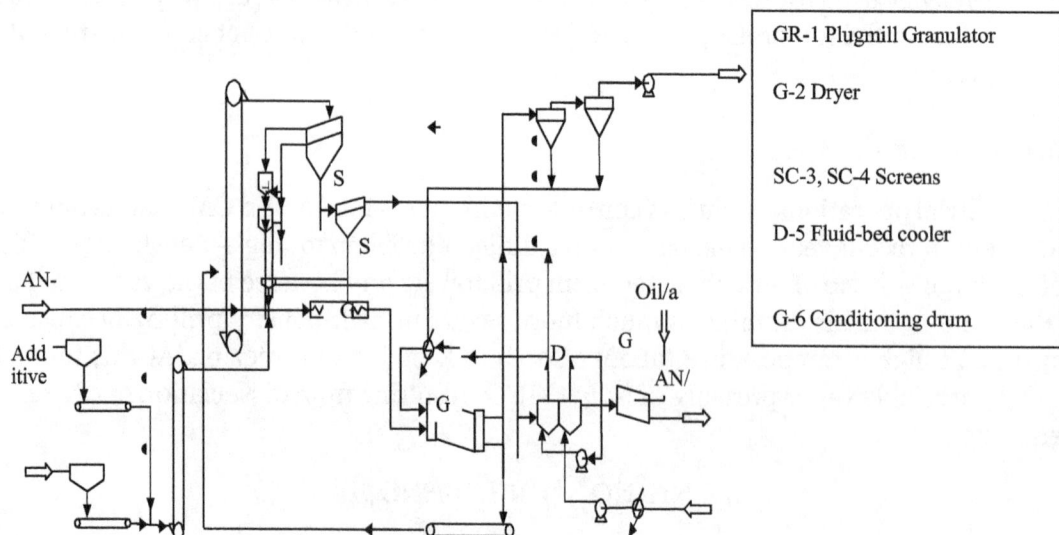

Pan granulation of CAN has proven difficult because the pan is very sensitive to heat and material balance factors. The product shape is irregular. The spherodizer® processes and all other processes need additives:

- Spherodizer®: ammonium sulfate, magnesium sulfate;

- Fluid bed: magnesium nitrate.

To improve product hardness, some manufacturers using the pugmill and ammonium sulfate – about 0.3%- 0.5% of SO_4.

The melt concentrations are also different. These concentrations are as follows:

- Fluid bed: 98%-99% wt.

- Pug-mill: 94.5%-95.5% wt.

- Drum: 93.5%-94.5% wt.

In the Hydro-Agri fluid bed granulation process, dedusting of the airstreams from the granulator and fluidized bed cooler is done by scrubbing with acidified weak AN solution. The scrubbing solution is mixed with the lime/AN mixture prior to final evaporation.

In pug-mill granulation, the AN melt and the lime are proportioned in ratio control to the pug-mill. Dust, undersize product, and crushed oversize are recycled to the pug-mill. The fresh, damp granules pass to a drying drum, and the granules are screened hot. For drying, the off-gas from the cooler is used. At full plant load the air heater for the dryer is turned off, and the plant operates auto-thermally. The drying air is dedusted in dry cyclones; final dedusting is performed in scrubbers. The on-size product is cooled in a fluidized bed cooler with conditioned air. Before storage or bagging the product is coated. Spilled product is returned to the granulation loop.

Ammonium Sulfate

Ammonium sulfate (American English; ammonium tetraoxosulfate (VI) is the IUPAC-recommended spelling; and ammonium sulphate in British English), $(NH_4)_2SO_4$, is an inorganic salt with a number of commercial uses. The most common use is as a soil fertilizer. It contains 21% nitrogen and 24% sulphur.

Uses

The primary use of ammonium sulfate is as a fertilizer for alkaline soils. In the soil the ammonium ion is released and forms a small amount of acid, lowering the pH balance of the soil, while contributing essential nitrogen for plant growth. The main disadvantage to the use of ammonium sulfate is its low nitrogen content relative to ammonium nitrate, which elevates transportation costs.

It is also used as an agricultural spray adjuvant for water-soluble insecticides, herbicides, and fungicides. There, it functions to bind iron and calcium cations that are present in both well water and plant cells. It is particularly effective as an adjuvant for 2,4-D (amine), glyphosate, and glufosinate herbicides.

Laboratory Use

Ammonium sulfate precipitation is a common method for protein purification by precipitation. As the ionic strength of a solution increases, the solubility of proteins in that solution decreases. Ammonium sulfate is extremely soluble in water due to its ionic nature, therefore it can "salt out" proteins by precipitation. Due to the high dielectric constant of water, the dissociated salt ions being cationic ammonium and anionic sulfate are readily solvated within hydration shells of water molecules. The significance of this substance in the purification of compounds stems from its ability to become more so hydrated compared to relatively more nonpolar molecules and so the desirable non polar molecules coalesce and precipitate out of the solution in a concentrated form. This method is called salting out and necessitates the use of high salt concentrations that can reliably dissolve in the aqueous mixture. The percentage of the salt used is in comparison to the maximal concentration of the salt the mixture can dissolve. As such, although high concentrations are needed for the method to work adding an abundance of the salt, over 100%, can also oversaturate the solution therefore contaminating the non polar precipitate with salt precipitate. A high salt concentration, which can be achieved by adding or increasing the concentration of ammonium sulfate in a solution, enables protein separation based on a decrease in protein solubility; this separation may be achieved by centrifugation. Precipitation by ammonium sulfate is a result of a reduction in solubility rather than protein denaturation, thus the precipitated protein can be solubilized through the use of standard buffers. Ammonium sulfate precipitation provides a convenient and simple means to fractionate complex protein mixtures.

In the analysis of rubber lattices, volatile fatty acids are analyzed by precipitating rubber with a 35% ammonium sulfate solution, which leaves a clear liquid from which volatile fatty acids are regenerated with sulfuric acid and then distilled with steam. Selective precipitation with ammonium sulfate, opposite to the usual precipitation technique which uses acetic acid, does not interfere with the determination of volatile fatty acids.

Food Additive

As a food additive, ammonium sulfate is considered generally recognized as safe (GRAS) by the U.S. Food and Drug Administration, and in the European Union it is designated by the E number E517. It is used as an acidity regulator in flours and breads.

Other Uses

Ammonium sulfate is used on a small scale in the preparation of other ammonium salts, especially ammonium persulfate.

Ammonium sulfate is listed as an ingredient for many United States vaccines per the Center for Disease Control.

A saturated solution of ammonium sulfate in heavy water (D_2O) is used as an external standard in sulfur (^{33}S) NMR spectroscopy with shift value of 0 ppm.

Ammonium sulfate has also been used in flame retardant compositions acting much like diammonium phosphate. As a flame retardant, it increases the combustion temperature of the material, decreases maximum weight loss rates, and causes an increase in the production of residue or char. Its flame retardant efficacy can be enhanced by blending it with ammonium sulfamate.

Ammonium sulfate has been used as a wood preservative, but due to its hygroscopic nature, this use has been largely discontinued because of associated problems with metal fastener corrosion, dimensional instability, and finish failures.

Preparation

Ammonium sulfate is made by treating ammonia, often as a by-product from coke ovens, with sulfuric acid:

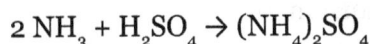

$$2\ NH_3 + H_2SO_4 \rightarrow (NH_4)_2SO_4$$

A mixture of ammonia gas and water vapor is introduced into a reactor that contains a saturated solution of ammonium sulfate and about 2 to 4% of free sulfuric acid at 60 °C. Concentrated sulfuric acid is added to keep the solution acidic, and to retain its level of free acid. The heat of reaction keeps reactor temperature at 60 °C. Dry, powdered ammonium sulfate may be formed by spraying sulfuric acid into a reaction chamber filled with ammonia gas. The heat of reaction evaporates all water present in the system, forming a powdery salt. Approximately 6000M tons were produced in 1981.

Ammonium sulfate also is manufactured from gypsum ($CaSO_4 \cdot 2H_2O$). Finely divided gypsum is added to an ammonium carbonate solution. Calcium carbonate precipitates as a solid, leaving ammonium sulfate in the solution.

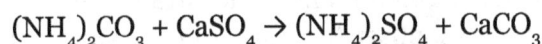

$$(NH_4)_2CO_3 + CaSO_4 \rightarrow (NH_4)_2SO_4 + CaCO_3$$

Ammonium sulfate occurs naturally as the rare mineral mascagnite in volcanic fumaroles and due to coal fires on some dumps.

Properties

Ammonium sulfate becomes ferroelectric at temperatures below -49.5 °C. At room temperature it crystallises in the orthorhombic system, with cell sizes of a = 7.729 Å, b = 10.560 Å, c = 5.951 Å. When chilled into the ferrorelectric state, the symmetry of the crystal changes to space group Pna2$_1$.

Reactions

Ammonium sulfate decomposes upon heating above 250 °C, first forming ammonium bisulfate. Heating at higher temperatures results in decomposition into ammonia, nitrogen, sulfur dioxide, and water.

As a salt of a strong acid (H_2SO_4) and weak base (NH_3), its solution is acidic; pH of 0.1 M solution is 5.5. In aqueous solution the reactions are those of NH_4^+ and SO_4^{-2} ions. For example, addition of barium chloride, precipitates out barium sulfate. The filtrate on evaporation yields ammonium chloride.

Ammonium sulfate forms many double salts (ammonium metal sulfates) when its solution is mixed with equimolar solutions of metal sulfates and the solution is slowly evaporated. With trivalent metal ions, alums such as ferric ammonium sulfate are formed. Double metal sulfates include ammonium cobaltous sulfate, ferrous diammonium sulfate, ammonium nickel sulfate which are known as Tutton's salts and ammonium ceric sulfate. Anhydrous double sulfates of ammonium also occur in the Langbeinites family.

Legislation and Control

In November 2009, a ban on ammonium sulfate, ammonium nitrate and calcium ammonium nitrate fertilizers was imposed in the former Malakand Division—comprising the Upper Dir, Lower Dir, Swat, Chitral and Malakand districts of the North West Frontier Province (NWFP) of Pakistan, by the NWFP government, following reports that they were used by militants to make explosives. In January 2010, these substances were also banned in Afghanistan for the same reason.

Production of Ammonium Sulphate

The majority of its production is coming from coking of coal as a byproduct. Ammonium sulphate is produced by the direct reaction of concentrated sulphuric acid and gaseous ammonia and proceeds according to the following steps.

1. Reaction of Ammonia and Sulphuric Acid:

Liquid ammonia is evaporated in an evaporator using 16 bar steam and preheated using low pressure steam.

The stiochiometric quantities of preheated gaseous ammonia and concentrated sulphuric acid (98.5% wt/wt) are introduced to the evaporator – crystalliser (operating under vacuum). These quantities are maintained by a flow recorder controller and properly mixed by a circulating pump (from upper part of the crystalliser to the evaporator).

2. Crystallization

The reaction takes place in the crystallizer where the generated heat of reaction causes evaporation of water making the solution supersaturated. The supersaturated solution settles down to the bottom of crystalliser where it is pumped to vacuum metallic filter where the A. S crystals are separated, while the mother liquor is recycled to the crystalliser.

3. Drying of the wet Ammonium Sulphate Crystals

The wet A.S crystals are conveyed (by belt conveyors) to the rotary dryer to be dried against hot air (steam heated) and then conveyed to the storage area where it naturally cooled and bagged.

Calcium Nitrate

As a fertilizer, calcium nitrate has special advantages for use on saline soils because the calcium displaces the sodium that is absorbed by clay in soils. For this reason, it may be preferred for use in areas with soil salinity problems. In addition, calcium nitrate has the advantage of being non acid forming; it improves the physical properties of exhausted and acidified soils and can be used as a topdressing.

Other applications include explosives, pyrotechnics and inorganic chemical operations. In some countries it is used I sizable tonnages as a de icing agent (at airports).

Properties of calcium nitrate:

The properties of pure calcium nitrate anhydride are given in the table. Calcium nitrate forms four modifications:

$-Ca(Na_3)_2$	
$-Ca(Na_3)_2 \cdot 2H_2O$	($H_2O = 1.8\%$)
$-Ca(Na_3)_2 \cdot 3H_2O$	($H_2O = 24.8\%$)
$-Ca(Na_3)_2 \cdot 4H_2O$	($H_2O = 30.5\%$)

Properties of calcium nitrate:

Formula	$Ca(NO_3)_2$
Appearance	White crystalline
Molecular weight	164.10
Melting point	555.7°C
Density 20°C	2.36(anhydride)
Solubility:	
At 0°C	50.5%
At 100°C	78.4%
At 150°C	79.0%
Critical relative humidity;	
At 20°C	54.8%
At 30°C	46.7%

By the addition of ammonium nitrate, the double salt $5Ca(NO_3)_2 \cdot NH_4NO_3 \cdot 10H_2O$ is formed. The melting point of the double salt is 100°-105°C; pH-value, 6-6.5.

Commercial Form, Storage and Transportation:

Calcium nitrate is traded as granulated, prilled or flaked products. The trend is toward granulated products. Typical specification of fertilizer grate is as follows.

Total nitrogen	15.5%
$Ca(NO_3)2$content	76.5 - 82%
NH_4NO_3 content	4.5 – 7.2%
H_2O content	12 - 17%

Hardness is a follows:

Prills	2.0 +/-0.5 kg/grain
Drum granules	3.0 +/-0.5 kg/grain
Pugmill granules	3.5 +/-0.5 kg/grain

Screen analysis is a follows:

Flakes	95% - 99% between 2 and 5 mm
Granules	90% - 92% between 2 and 4 mm
Prills	93% - 95% between 1 and 3 mm

Bulk storage is to be avoided at all costs. Immediate bagging in multilayered bituminoid paper bags,. PE-lined jute bags, or mono layer PE bags is mandatory.

As with other oxidizing nitrate fertilizers, precautions should be taken to avoid impregnation with organic material and contact with a heat source. Special storage and shipping regulations must be observed.

Calcium nitrate is produced by dissolving the calcium carbonate (lime stone) with nitric acid, according to the following reaction:

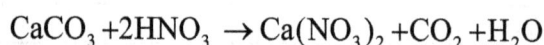

$$CaCO_3 + 2HNO_3 \rightarrow Ca(NO_3)_2 + CO_2 + H_2O$$

The lime stone is transported to the site as small size stones and lifted to the dissolving tower. The nitric acid is fed to the bottom of the dissolving tower and the formed calcium nitrate is fed to the settling tank. After settling, the excess acid is neutralized with ammonia. The nitrogen content is adjusted with ammonium nitrate. The fertilizer is produced in the liquid state and the nitrogen content of the final product is adjusted to the required specifications using ammonium nitrate.

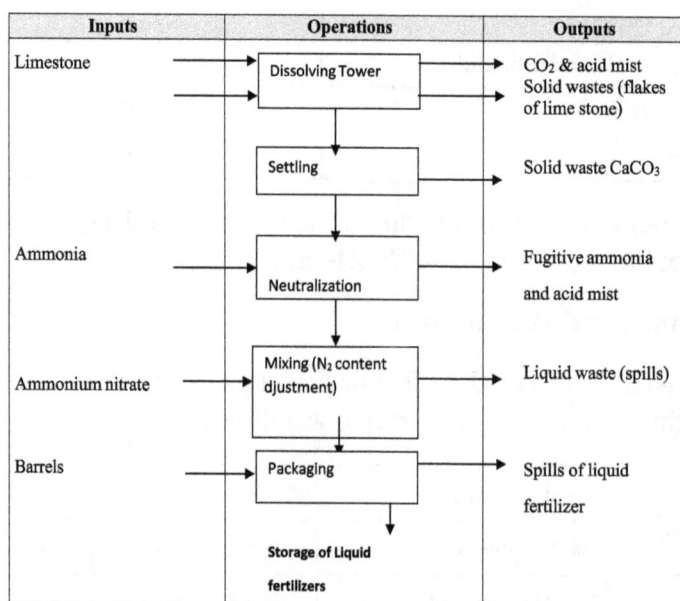

Inputs	Operations	Outputs
Limestone	Dissolving Tower	CO₂ & acid mist Solid wastes (flakes of lime stone)
	Settling	Solid waste CaCO₃
Ammonia	Neutralization	Fugitive ammonia and acid mist
Ammonium nitrate	Mixing (N₂ content djustment)	Liquid waste (spills)
Barrels	Packaging	Spills of liquid fertilizer
	Storage of Liquid fertilizers	

Process Flow Diagram for Calcium Nitrate

Ammonium Chloride

Ammonium chloride ride is used fertilization either such or in a variety of compound fertilizers. Examples are:

18-22-0(ammonium phosphate chloride)

16-0-20(ammonium-potassium chloride)

14-14-14

12-18-14

Ammonium chloride is used in other grades of compound fertilizers in combination with urea or ammonium sulfate. Advantages of ammonium chloride are that it has a higher concentration than ammonium sulfate and a somewhat lower cost per unit of N (in Japan). It has some agronomic advantages for rice; nitrification is less rapid than with urea or ammonium sulfate and, there-fore N losses are lower and yields are higher.

Although ammonium chloride is best known as a rice fertilizer, it has been successfully tested and used on other crops such as wheat, barley, sugarcane, maize, fiber crops, and sorghum in a variety of climatic conditions.

Ammonium chloride is as highly acid-forming as ammonium sulfate per unit of N, which can be a disadvantage. Other disadvantages are its low N content compared with urea or ammonium nitrate and the high chloride content, which can be harmful on some crops or soils. Nevertheless, it is possible that ammonium chloride fertilizer could become a useful outlet for surplus chlorine or byproduct hydrochloric acid that arises from time to time.

Another ruse full feature of ammonium chloride is that it can be applied to rice with safety in the presence of certain fungi, which would reduce ammonium sulfate to toxic sulfides.

Properties of ammonium chloride- The properties of ammonium chloride are given in table.

Commercial form, storage, and transportation- The fertilizer-grade product contains 25% N. The product can be in the form of either crystals or granules. Coarse crystalline or granular forms are preferred for direct application, whereas fine crystals can be used in compound fertilizers.

Raw Materials- principal raw materials are common salt (Naclo and anhydrous ammonia in the case of the dual-salt process or anhydrous ammonia and hydrochloric acid (HCl) for the direct-neutralization method. To take advantage of byproduct CO_2, it is advisable to install the dual-salt process at a site where anhydrous ammonia is produced. For the direct

Properties of Ammonium chloride	
Formula	NH_4Cl
Molecular weight	53.5
Nitrogen content	26%

Color	white
Density of solid, 20°C	1.526

Solubility, g/100g of water at:

Temperature, °C	
1	29.4
20	37.2
40	45.8
60	55.3
80	65.6
100	77.3
115.6 (boiling point)	87.3

Effect of heat: Ammonium chloride begins to dissociate at 350°C and sublimes at 520 °C

Crystal relative humidity

At 20 °C   79.2

At 30 °C   77.5

Neutralization method, byproduct HCl may be used, e.g., byproduct HCl from the production of potassium sulfate by the Mannheim process.

Production Methods- Several methods for producing ammonium chloride are used: the order of importance is follows:

1. The dual-salt process, whereby ammonium chloride and sodium carbonate are produced simultaneously.

2. Direct neutralization of ammonia with hydrochloric acid.

The Dual-Salt Process- Most ammonium chloride used in India, china, and Japan for fertilizer purposes is produced by the dual-salt process as shown.

In this method, ammonium chloride is salted out by the addition of solid, washed sodium chloride rather than decomposition by lime liquor to recover ammonium carbonate:

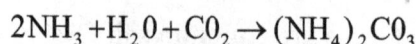

$$2NH_3 + H_2O + CO_2 \rightarrow (NH_4)_2CO_3$$

Additional carbonation produces ammonium bicarbonate

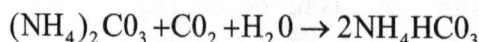

$$(NH_4)_2CO_3 + CO_2 + H_2O \rightarrow 2NH_4HCO_3$$

The addition of sodium chloride yields sodium bicarbonate and ammonium chloride:

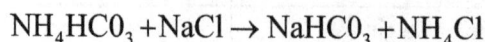

$$NH_4HCO_3 + NaCl \rightarrow NaHCO_3 + NH_4Cl$$

The sodium bicarbonate is separated by centrifuging or filtration and calcined to produce sodium carbonate and CO_2; the latter is recycled to the system.

The Direct-neutralization Method- Ammonium chloride of high purity can be made by the direct reaction between anhydrous ammonia vapor and hydrochloric acid gas, according to the reaction

$$NH_{3(g)} + HCl_{(g)} \rightarrow NH_4Cl$$

The reaction is exothermic (42,000cal/g-mole). In most cases, neutralization is undertaken at reduced pressures of 250-300mm of mercury in one or more rubber-lined steel vacuum reaction vessels protected with an additional inner lining of inert brick. Concentrated hydrochloric acid gas is passed through an aspirator, wherein it is diluted with air to about 20% concentration and enters the reaction vessel via a vertical sparser tube. According to preference, ammonia gas is introduced either by a second sparser or by tangential nozzles in the base of the reaction vessel. Agitation is provided by the large volume of air entering the reactor with the hydrochloric acid vapor; thus, the need for a mechanical agitator with its additional power requirements and maintenance problems is avoided.

Similarly, operation under vacuum not only provides excellent cooling but simultaneously prevents escape of noxious vapors and eliminates the need for hydrochloric acid-vapor blowers, plus their attendant cost and maintenance charges. A reduced pressure of 250- 330mm of mercury and a corresponding slurry temperature of 75°-80°C represent typical operating conditions.

Mother liquor the centrifuge is pumped back to the saturator(s) via a storage tank. Saturator off gases must be well scrubbed before entering the vacuum pump or ejector unit to prevent corrosioon and to eliminate air pollution. A two-stage scrubbing system is usually employed and may consist of a direct, barometric scrubber condenser followed by a wetted, packed tower. Liquor from the scrubber-condenser is returned to the mother liquor tank and is evaporated in the saturator, thus providing a means of temperature control and recovery.

As with other processes involving reactions between hydrochloric acid (or chlorides) and ammonia, traces of free chlorine in the acid feed can lead to disastrous explosions caused by the formation of nitrogen trichloride in the saturator. Hence, adequate safety precautions must be installed whereby the HCl gas feed is monitored and the flow discontinued when chlorine is detected. This can be accomplished by such means as bypassing a small stream of gas through a photocell- calorimeter unit containing potassium iodide or using a modern continuous gas analyzer of the absorption or chromatographic type.

After separation and drying, the crystalline ammonium chloride is bagged as quickly as possible to minimize subsequent storage and application difficulties.

Urea

Urea or carbamide is an organic compound with the chemical formula $CO(NH_2)_2$. The molecule has two $-NH_2$ groups joined by a carbonyl (C=O) functional group.

Urea serves an important role in the metabolism of nitrogen-containing compounds by animals and is the main nitrogen-containing substance in the urine of mammals. It is solid, colourless, and odorless (although the ammonia that it gives off in the presence of water, including water vapor in the air, has a strong odor). It is highly soluble in water and practically non-toxic (LD50 is 15 g/kg for rat). Dissolved in water, it is neither acidic nor alkaline. The body uses it in many processes, the most notable one being nitrogen excretion. Urea is widely used in fertilizers as a convenient source of nitrogen. Urea is also an important raw material for the chemical industry. The synthesis of this organic compound by Friedrich Wöhler in 1828 from an inorganic precursor was an important milestone in the development of organic chemistry, as it showed for the first time that a molecule found in living organisms could be synthesized in the lab without biological starting materials (thus contradicting a theory widely prevalent at one time, called vitalism).

More than 90% of world production of urea is destined for use as a nitrogen-release fertilizer. Urea has the highest nitrogen content of all solid nitrogenous fertilizers in common use. Therefore, it has the lowest transportation costs per unit of nitrogen nutrient. The standard crop-nutrient rating of urea is 46-0-0.

Many soil bacteria possess the enzyme urease, which catalyzes the conversion of the urea molecule to two ammonia molecules and one carbon dioxide molecule, thus urea fertilizers are very rapidly transformed to the ammonium form in soils. Among soil bacteria known to carry urease, some ammonia-oxidizing bacteria (AOB) such as species of Nitrosomonas are also able to assimilate the carbon dioxide released by the reaction to make biomass via the Calvin Cycle, and harvest energy by oxidizing ammonia (the other product of urease) to nitrite, a process termed nitrification. Nitrite-oxidizing bacteria, especially Nitrobacter, oxidize nitrite to nitrate, which is extremely mobile in soils and is a major cause of water pollution from agriculture. Ammonia and nitrate are readily absorbed by plants, and are the dominant sources of nitrogen for plant growth. Urea is also used in many multi-component solid fertilizer formulations. Urea is highly soluble in water and is, therefore, also very suitable for use in fertilizer solutions (in combination with ammonium nitrate: UAN), e.g., in 'foliar feed' fertilizers. For fertilizer use, granules are preferred over prills because of their narrower particle size distribution, which is an advantage for mechanical application.

The most common impurity of synthetic urea is biuret, which impairs plant growth.

Urea is usually spread at rates of between 40 and 300 kg/ha but rates vary. Smaller applications incur lower losses due to leaching. During summer, urea is often spread just before or during rain to minimize losses from volatilization (process wherein nitrogen is lost to the atmosphere as ammonia gas). Urea is not compatible with other fertilizers.

Because of the high nitrogen concentration in urea, it is very important to achieve an even spread. The application equipment must be correctly calibrated and properly used. Drilling must not occur on contact with or close to seed, due to the risk of germination damage. Urea dissolves in water for application as a spray or through irrigation systems.

In grain and cotton crops, urea is often applied at the time of the last cultivation before planting. In high rainfall areas and on sandy soils (where nitrogen can be lost through leaching) and where good in-season rainfall is expected, urea can be side- or top-dressed during the growing season.

Top-dressing is also popular on pasture and forage crops. In cultivating sugarcane, urea is side-dressed after planting, and applied to each ratoon crop.

In irrigated crops, urea can be applied dry to the soil, or dissolved and applied through the irrigation water. Urea will dissolve in its own weight in water, but it becomes increasingly difficult to dissolve as the concentration increases. Dissolving urea in water is endothermic, causing the temperature of the solution to fall when urea dissolves. As a practical guide, when preparing urea solutions for fertigation (injection into irrigation lines), dissolve no more than 30 kg urea per 100 L water.

In foliar sprays, urea concentrations of 0.5% – 2.0% are often used in horticultural crops. Low-biuret grades of urea are often indicated.

Urea absorbs moisture from the atmosphere and therefore is typically stored either in closed/sealed bags on pallets or, if stored in bulk, under cover with a tarpaulin. As with most solid fertilizers, storage in a cool, dry, well-ventilated area is recommended.

Urea Production Processes

The commercial synthesis of urea involves the combination of ammonia and carbon dioxide at high pressure to form ammonium carbamate which is subsequently dehydrated by the application of heat to form urea and water.

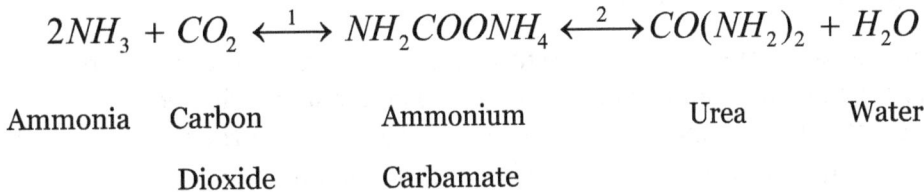

$$2NH_3 + CO_2 \xleftrightarrow{1} NH_2COONH_4 \xleftrightarrow{2} CO(NH_2)_2 + H_2O$$

| Ammonia | Carbon Dioxide | Ammonium Carbamate | Urea | Water |

Reaction 1 is fast and exothermic and essentially goes to completion under the reaction conditions used industrially. Reaction 2 is slower and endothermic and does not go to completion.

The conversion (on a CO_2 basis) is usually in the order of 50-80%. The conversion increases with increasing temperature and NH_3/CO_2 ratio and decreases with increasing H_2O/CO_2 ratio.

The design of commercial processes has involved the consideration of how to separate the urea from the other constituents, how to recover excess NH_3 and decompose the carbamate for recycle. Attention was also devoted to developing materials to withstand the corrosive carbamate solution and to optimise the heat and energy balances.

The simplest way to decompose the carbamate to CO_2 and NH_3 requires the reactor effluent to be depressurised and heated. The earliest urea plants operated on a "Once Through" principle where the off-gases were used as feedstocks for other products.

Subsequently "Partial Recycle" techniques were developed to recover and recycle some of the NH_3 and CO_2 to the process. It was essential to recover all of the gases for recycle to the synthesis to optimise raw material utilisation and since recompression was too expensive an alternative method was developed. This involved cooling the gases and re- combining them to form carbamate liquor which was pumped back to the synthesis. A series of loops involving carbamate decomposers at progressively lower pressures and carbamate condensers were used. This was known as the "Total Recycle Process". A basic consequence of recycling the gases was that the NH_3/CO_2 molar ratio in the reactor increased thereby increasing the urea yield.

Significant improvements were subsequently achieved by decomposing the carbamate in the reactor effluent without reducing the system pressure. This "Stripping Process" dominated synthesis technology and provided capital/energy savings. Two commercial stripping systems were developed, one using CO_2 and the other using NH_3 as the stripping gases.

Since the base patents on stripping technology have expired, other processes have emerged which combine the best features of Total Recycle and Stripping Technologies. For convenience total recycle processes were identified as either "conventional" or "stripping" processes.

The urea solution arising from the synthesis/recycle stages of the process is subsequently concentrated to a urea melt for conversion to a solid prilled or granular product. Improvements in process technology have concentrated on reducing production costs and minimising the environmental impact. These included boosting CO_2 conversion efficiency, increasing heat recovery, reducing utilities consumption and recovering residual NH_3 and urea from plant effluents. Simultaneously the size limitation of prills and concern about the prill tower off-gas effluent were responsible for increased interest in melt granulation processes and prill tower emission abatement. Some or all of these improvements have been used in updating existing plants and some plants have added computerised systems for process control.

New urea installations vary in size from 800 to 2,000t.d-1 and typically would be 1,500t.d-1 units. Modern processes have very similar energy requirements and nearly 100% material efficiency. There are some differences in the detail of the energy balances but they are deemed to be minor in effect. Block flow diagrams for CO_2 and NH_3 stripping total recycle processes are shown in the Block Diagram figures.

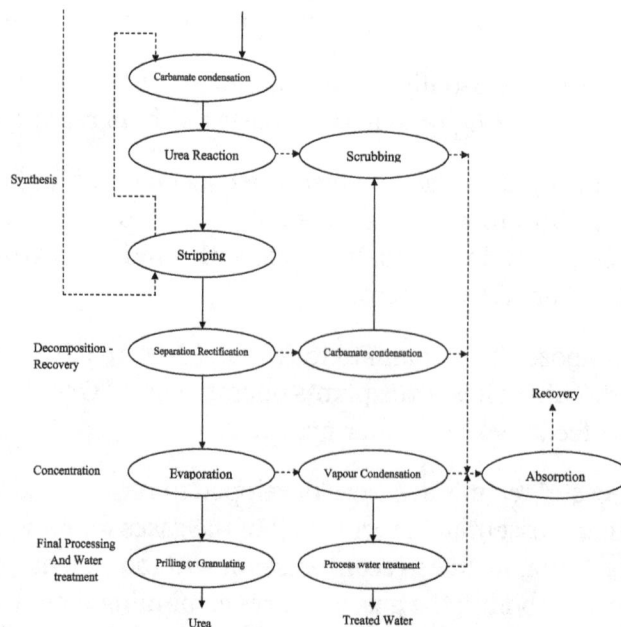

Block Diagram of a Total Recycle CO2 Stripping Urea Process.

The process water from each process discussed in this chapter is purified by recovery of dissolved urea, NH_3 and CO_2 which are recycled to the synthesis section via a low pressure carbamate condensation system.

Carbon Dioxide Stripping Process

NH3 and CO_2 are converted to urea via ammonium carbamate at a pressure of approximately 140bar and a temperature of 180-185°C. The molar NH_3/CO_2 ratio applied in the reactor is 2.95. This results in a CO_2 conversion of about 60% and an NH_3 conversion of 41%.

The reactor effluent, containing unconverted NH_3 and CO_2 is subjected to a stripping operation at essentially reactor pressure, using CO_2 as stripping agent. The stripped off NH_3 and CO_2 are then partially condensed and recycled to the reactor. The heat evolving from this condensation is used to produce 4.5bar steam some of which can be used for heating purposes in the downstream sections of the plant. Surplus 4.5bar steam is sent to the turbine of the CO_2 compressor.

The NH_3 and CO_2 in the stripper effluent are vaporised in a 4bar decomposition stage and subsequently condensed to form a carbamate solution, which is recycled to the 140bar synthesis section. Further concentration of the urea solution leaving the 4bar decomposition stage takes place in the evaporation section, where a 99.7% urea melt is produced.

The urea production processes are continued below:

Ammonia Stripping Process

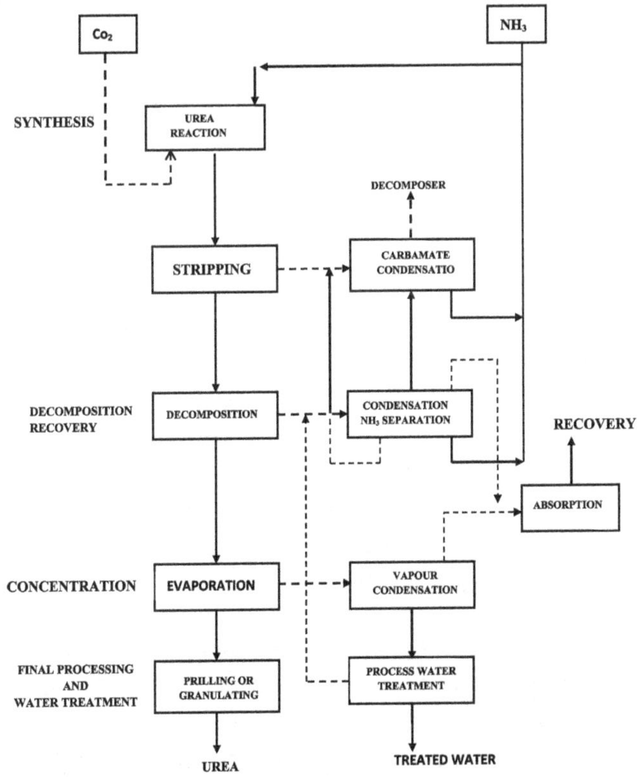

Block Diagram of a Total Recycle NH_3 Stripping Urea Process

NH_3 and CO_2 are converted to urea via ammonium carbamate at a pressure of 150bar and a temperature of 180°C. A molar ratio of 3.5 is used in the reactor giving a CO_2 conversion of 65%. The reactor effluent enters the stripper where a large part of the unconverted carbamate is decomposed by the stripping action of the excess NH_3. Residual carbamate and CO_2 are recovered downstream of the stripper in two successive stages operating at 17 and 3.5bar respectively. NH_3 and CO_2 vapours from the stripper top are mixed with the recovered carbamate solution from the High Pressure (HP)/Low Pressure (LP) sections, condensed in the HP carbamate condenser and fed to the reactor. The heat of condensation is used to produce LP steam. The urea solution leaving the LP decomposition stage is concentrated in the evaporation section to a urea melt.

Advanced Cost & Energy Saving (ACES) Process

In this process the synthesis section operates at 175bar with an NH_3/CO_2 molar ratio of 4 and a temperature of 185 to 190°C. The reactor effluent is stripped at essentially reactor pressure using CO_2 as the stripping agent. The overhead gas mixture from the stripper is fed to two carbamate condensers in parallel where the gases are condensed and recycled under gravity to the reactor along with absorbent solutions from the HP scrubber and absorber. The heat generated in the first carbamate condenser is used to generate 5bar steam and the heat formed in the second condenser is used to heat the solution leaving the stripper bottom after pressure reduction. The inerts in the synthesis section are purged to the scrubber from the reactor top for recovery and recycle of NH_3 and CO_2. The urea solution leaving the bottom of the stripper is further purified in HP and LP decomposer operating at approx. 17.5bar and 2.5bar respectively. The separated NH_3 and CO_2 are recovered to the synthesis via HP and LP absorbers. The aqueous urea solution is first concentrated to 88.7%wt in a vacuum concentrator and then to the required concentration for prilling or granulating.

Isobaric Double Recycle (IDR) Process

In this process the urea synthesis takes place at 180-200bar and 185-190°C. The NH_3/CO_2 ratio is approximately 3.5-4, giving about 70% CO_2 conversion per pass. Most of the unconverted material in the urea solution leaving the reactor is separated by heating and stripping at synthesis pressure using two strippers, heated by 25bar steam, arranged in series. In the first stripper, carbamate is decomposed/stripped by ammonia and the remaining ammonia is removed in the second stripper using carbon dioxides as stripping agent. Whereas all the carbon dioxide is fed to the plant through the second stripper, only 40% of the ammonia is fed to the first stripper. The remainder goes directly to the reactor for temperature control. The ammonia-rich vapours from the first stripper are fed directly to the urea reactor. The carbon dioxide-rich vapours from the second stripper are recycled to the reactor via the carbamate condenser, irrigated with carbamate solution recycled from the lower-pressure section of the plant.

The heat of condensation is recovered as 7bar steam which is used down-stream in the process. Urea solution leaving the IDR loop contains unconverted ammonia, carbon dioxide and carbamate. These residuals are decomposed and vaporised in two successive distillers, heated with low pressure recovered steam. After this, the vapours are condensed to carbamate solution and recycled to the synthesis loop. The urea solution leaving the LP decomposition for further concentration, is fed to two vacuum evaporators in series, producing the urea melt for prilling and granulation.

Process Water Sources and Quantities

A 1,000t.d-1 urea plant generates on average approximately 500m3.d-1 process water containing 6% NH_3, 4% CO_2 and 1.0% urea (by weight). The principal source of this water is the synthesis reaction where 0.3tonnes of water is formed per tonne of urea e.g.

$$2NH_3 + CO_2 \; CO(NH_2)_2 + H_2O$$

The other sources of water are ejector steam, flush and seal water and steam used in the waste water treatment plant.

The principal sources of urea, NH_3 and CO_2 in the process water are:-

– Evaporator condensate

The NH_3 and urea in the evaporator condensate are attributable to:-

– The presence of NH_3 in the urea solution feed to the evaporator

– The formation of biuret and the hydrolysis of urea in the evaporators, both reactions liberating NH_3

$$2CO(NH_2)_2 \; H_2NCONHCONH_2 + NH_3$$

$$CO(NH_2)_2 + H_2O \; 2NH_3 + CO_2$$

– Direct carry over of urea from the evaporator separators to the condensers (physical entrainment)

– The formation of NH_3 from the decomposition of urea to isocyanic acid $CO(NH_2)_2 \; HNCO + NH_3$

The reverse reaction occurs on cooling the products in the condensers

– Off-gases from the recovery/recirculation stage absorbed in the process water

– Off-gases from the synthesis section absorbed in the process water

– Flush and purge water from pumps

– Liquid drains from the recovery section

The purpose of the water treatment is to remove NH_3, CO_2 and urea from the process water and recycle the gases to the synthesis. This ensures raw material utilisation is optimised and effluent is minimised.

Prilling and Granulation

In urea fertilizer production operations, the final product is in either prilled or granular form. Production of either form from urea melt requires the use of a large volume of cooling air which is subsequently discharged to the atmosphere.

Prilling

The concentrated (99.7%) urea melt is fed to the prilling device (e.g. rotating bucket/shower type spray head) located at the top of the prilling tower. Liquid droplets are formed which solidify and cool on free fall through the tower against a forced or natural up- draft of ambient air. The product is removed from the tower base to a conveyor belt using a rotating rake, a fluidised bed or a conical hopper. Cooling to ambient temperature and screening may be used before the product is finally transferred to storage.

The design/operation of the prilling device exerts a major influence on product size. Collision of the molten droplets with the tower wall as well as inter-droplet contact causing agglomeration must be prevented. Normally mean prill diameters range from 1.6-2.0mm for prilling operations. Conditioning of the urea melt and "crystal seeding" of the melt, may be used to enhance the anti-caking and mechanical properties of the prilled product during storage/ handling.

Granulation

Depending on the process a 95-99.7% urea feedstock is used. The lower feedstock concentration allows the second step of the evaporation process to be omitted and also simplifies the process condensate treatment step. The basic principle of the process involves the spraying of the melt onto recycled seed particles or prills circulating in the granulator. A slow increase in granule size and drying of the product takes place simultaneously. Air passing through the granulator solidifies the melt deposited on the seed material. Processes using low concentration feedstock require less cooling air since the evaporation of the additional water dissipates part of the heat which is released when the urea crystallizes from liquid to solid.

All the commercial processes available are characterised by product recycle, and the ratio of recycled to final product varies between 0.5 and 2.5. Prill granulation or fattening systems have a very small recycle, typically 2 to 4%. Usually the product leaving the granulator is cooled and screened prior to transfer to storage. Conditioning of the urea melt prior to spraying may also be used to enhance the storage/handling characteristics of the granular product.

Urea: Safety

Urea can be irritating to skin, eyes, and the respiratory tract. Repeated or prolonged contact with urea in fertilizer form on the skin may cause dermatitis.

High concentrations in the blood can be damaging. Ingestion of low concentrations of urea, such as are found in typical human urine, are not dangerous with additional water ingestion within a reasonable time-frame. Many animals (e.g., dogs) have a much more concentrated urine and it contains a higher urea amount than normal human urine; this can prove dangerous as a source of liquids for consumption in a life-threatening situation (such as in a desert).

Urea can cause algal blooms to produce toxins, and its presence in the runoff from fertilized land may play a role in the increase of toxic blooms.

The substance decomposes on heating above melting point, producing toxic gases, and reacts violently with strong oxidants, nitrites, inorganic chlorides, chlorites and perchlorates, causing fire and explosion.

References

- Catherine E. Housecroft; Alan G. Sharpe (2008). "Chapter 15: The group 15 elements". Inorganic Chemistry, 3rd Edition. Pearson. ISBN 978-0-13-175553-6

- "Select Committee on GRAS Substances (SCOGS) Opinion: Ammonium sulfate". U.S. Food and Drug Administration. August 16, 2011. Retrieved March 2, 2013

- Luzzati, V. (1951). "Structure cristalline de l'acide nitrique anhydre". Acta Crystallographica (in French). 4 (2): 120–131. doi:10.1107/S0365110X51000404

- Considine, Douglas M., ed. (1974). Chemical and process technology encyclopedia. New York: McGraw-Hill. pp. 769–72. ISBN 978-0-07-012423-3

- "Official Subway Restaurants U.S. Products Ingredients Guide". Archived from the original on August 14, 2011. Retrieved March 2, 2013

- O'Neal, C. L.; Crouch, D. J.; Fatah, A. A. (2000). "Validation of twelve chemical spot tests for the detection of drugs of abuse". Forensic Science International. 109 (3): 189–201. PMID 10725655. doi:10.1016/S0379-0738(99)00235-2

- Clesceri, Lenore S.; Greenberg, Arnold E.; Eaton, Andrew D., eds. (1998). Standard methods for the examination of water and wastewater (20th ed.). American Public Health Association, American Water Works Association, Water Environment Federation. ISBN 978-0-87553-235-6

- "PAKISTAN: 'Anti-terrorist' fertilizer ban hinders farmers". IRIN Humanitarian News and Analysis. 2010. Retrieved April 24, 2013

- Karl-Heinz Zapp "Ammonium Compounds" in Ullmann's Encyclopedia of Industrial Chemistry 2012, Wiley-VCH, Weinheim. doi:10.1002/14356007.a02_243

- Lide, David R., ed. (2006). CRC Handbook of Chemistry and Physics (87th ed.). Boca Raton, FL: CRC Press. ISBN 0-8493-0487-3

Techniques for Phosphorus Fertilizers Production

Phosphorous is a macronutrient and an important fertilizer. Phosphate rock is the main source of phosphate. The chapter closely examines the key concepts of phosphorus fertilizers to provide an extensive understanding of the subject.

Sulphuric Acid

More sulphuric acid is produced than any other chemical in the world. The output of sulphuric acid at base metal smelters today represents about 20% of all acid production. Most of its uses are actually indirect in that the sulphuric acid is used as a reagent rather than an ingredient. The largest single sulphuric acid consumer by far is the fertiliser industry. Sulphuric acid is used with phosphate rock in the manufacture of phosphate fertilisers. Smaller amounts are used in the production of ammonium and potassium sulphate. Substantial quantities are used as an acidic dehydrating agent in organic chemical and petrochemical processes, as well as in oil refining. In the metal processing industry, sulphuric acid is used for pickling and descaling steel; for the extraction of copper, uranium and vanadium from ores; and in non-ferrous metal purification and plating. In the inorganic chemical industry, it is used most notably in the production of titanium dioxide.

Certain wood pulping processes for paper also require sulphuric acid, as do some textile and fibres processes (such as rayon and cellulose manufacture) and leather tanning. Other end uses for sulphuric acid include: effluent/water treatment, plasticisers, dyestuffs, explosives, silicate for toothpaste, adhesives, rubbers, edible oils, lubricants and the manufacture of food acids such as citric acid and lactic acid. Probably the largest use of sulphuric acid in which this chemical becomes incorporated into the final product is in organic sulphonation processes, particularly for the production of detergents. Many pharmaceuticals are also made by sulphonation processes.

Production of Sulphuric Acid

Liquid sulphur is a product of the desulphurisation of natural gas and crude oil by the Claus- Process, with the cleaning of coal flue gas as a second source. The third way is the melting of natural solid sulphur (Frash-process) but this is not in frequent use because there are many difficulties in removing the contaminants. The following is a typical analysis of molten sulphur (quality: bright yellow):

Ash max. 0.015% weight

Carbon max. 0.02% weight

Hydrogen sulphide ca. 1-2mg.kg-1

Sulphur dioxide 0mg.kg-1

Arsenic max. 1mg.kg-1

Mercury max. 1mg.kg-1

Water max. 0.05% weight

Liquid sulphur is transported in ships, railcars and trucks made of mild steel. Special equipment is used for all loading and unloading facilities. Liquid sulphur is stored in insulated and steam heated mild steel tanks. The tank is is equipped with submerged fill lines to avoid static charges and reduce agitation in the tank. The ventilation of the tanks is conventionally free. A further fact is less de-gasing of hydrogen sulphide and sulphur dioxide. All pipes and pumps are insulated and steam heated. The normal temperature level of the storage and handling is about 125-145°C.

Material Processing

Conversion of SO$_2$ into SO$_3$

The design and operation of sulphuric acid plants are focused on the following gas phase chemical equilibrium reaction with a catalyst:-

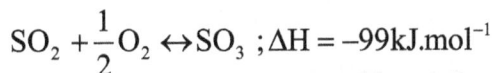

$$SO_2 + \frac{1}{2}O_2 \leftrightarrow SO_3 \; ; \Delta H = -99kJ.mol^{-1}$$

This reaction is characterised by the conversion rate, which is defined as follows:- conversion rate =100 (%)

Both thermodynamic and stoichiometric considerations are taken into account in maximizing the formation of SO$_3$. The Lechatelier-Braun Principle is usually taken into account in deciding how to optimise the equilibrium. This states that when an equilibration system is subjected to stress, the system will tend to adjust itself in such a way that part of the stress is relieved.

For SO$_2$/SO$_3$ systems, the following methods are available to maximise the formation of SO$_3$:-

– Removal of heat – a decrease in temperature will favour the formation of SO$_3$ since this is an exothermic process

– Increased oxygen concentration

– Removal of SO$_3$ (as in the case of the double absorption process)

– Raised system pressure

– Selection of the catalyst to reduce the working temperature (equilibrium)

– Increased reaction time

Optimum overall system behaviour requires a balance between reaction velocity and equilibrium.

However, this optimum also depends on the SO_2 concentration in the raw gas and on its variability with time. Consequently, each process is more or less specific for a particular SO_2 source.

Absorption of SO_3

Sulphuric acid is obtained from the absorption of SO_3 and water into H_2SO_4 (with a concentration of at least 98%).

The efficiency of the absorption step is related to:-

– The H_2SO_4 concentration of the absorbing liquid (98.5-99.5%)

– The range of temperature of the liquid (normally 70°C-120°C)

– The technique of the distribution of acid

– The raw gas humidity (mist passes the absorption equipment)

– The mist filter

– The temperature of incoming gas

The co-current or counter-current character of the gas stream in the absorbing liquid SO_3 emissions depend on:-

– The temperature of gas leaving absorption

– The construction and operation of the final absorber

– The device for separating H_2SO_4 aerosols

– The acid mist formed upstream of the absorber through the presence of water vapour

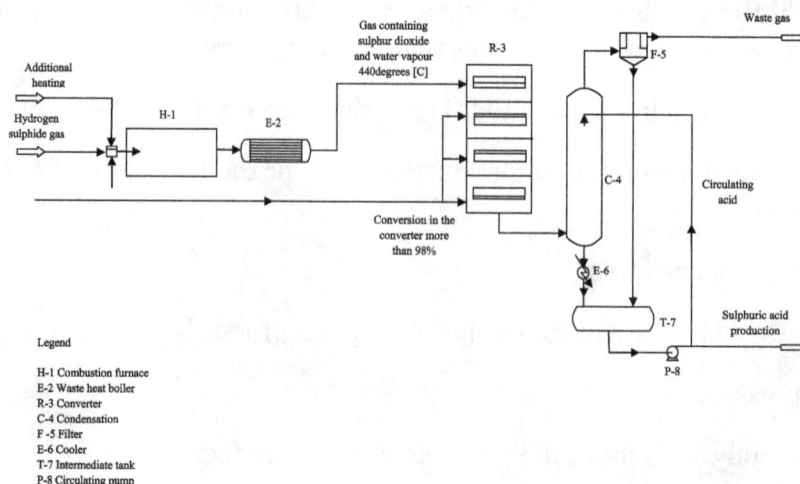

A typical sulphuric acid production Plant

Product Finishing

Dilution of Absorber Acids

The acid produced, normally 95.5%-96.5% or 98.5%-99.5%, is diluted with water or steam condensate down to the commercial concentrations: 25%, 37%, 48%, 78%, 96% and 98% H_2SO_4. The dilution can be made in a batch process or continuously through in-line mixing.

SO_2- Stripping

A small amount of air is blown through the warm acid in a column or tower to reduce the remaining SO_2 in the acid to < 20mg SO_2 kg-1. The air containing SO_2 is returned to the process.

Purification

Sulphuric acid from the start up of acid plants after long repair may be contaminated and clouded by insoluble iron sulphate, or silicate from bricks or packing. The acid can be filtered using conventional methods. Filter elements are required in the filling lines for tanker or railway loading to maintain quality.

Denitrification

A number of different methods are used for the denitrification of sulfuric acid and oleum. Various chemicals are used to reduce nitrosylsulphuric acid ($NOHSO_4$) or nitrate to N_2 or NxOy. The reactant must be added in stoichiometric amounts.

Decolourisation

Acid produced from smelter plants or from acid regeneration plants can contain hydrocarbons or carbonaceous material, which is absorbed in sulphuric acid. This causes a 'black' colour. The decolourisation is known as "acid bleaching".

Use of Auxiliary Chemicals/Materials Catalysts

When producing sulphuric acid by the contact process an important step is to produce sulphur trioxide by passing a gas mixture of sulphur dioxide and oxygen over a catalyst according to the equation:-

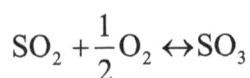

$$SO_2 + \frac{1}{2}O_2 \leftrightarrow SO_3$$

Without a catalyst this reaction needs a very high temperature to have a realistic rate. The equilibrium is however in favour of SO_2 – formation at higher temperatures which makes the conversion very poor. Of all substances tested for catalytic activity toward sulphur dioxide oxidation only vanadium compounds, platinum and iron oxide have proven to be technically satisfactory. Today vanadium pentoxide is used almost exclusively. Commercial catalysts contain 4-9 wt % vanadium pentoxide, V_2O_5, as the active component, together with alkali metal sulphate promoters. Under operating conditions these form a liquid melt in which the reaction is thought to take place. Normally potassium sulphate is used as a promoter but in recent years also caesium sulphate has been used. Caesium sulphate lowers the melting point, which means that the catalyst can be used at lower temperatures. The carrier material is silica in different forms. The catalyst components are mixed together to form a paste and then usually extruded into solid cylindrical pellets, rings or star-rings which are then baked at elevated temperatures. Ring (or star-ring) shaped catalysts, which are mostly used today, give a lower pressure drop and are less sensitive to dust blockage. The lower temperature limit is 410-430°C for conventional catalysts and 380-390°C for caesium doped catalysts. The upper temperature limit is 600-650°C above which, catalytic activity can be lost permanently due to reduction of the internal surface. The average service life for the catalyst is about 10 years. Service life is generally determined by catalyst losses during the screening of the catalyst which is necessary from time to time to remove dust.

Intermediate and Final Product Storage

There is no air pollution problem connected with the storage, handling and shipping of sulphuric acid because of the very low vapour pressure of H_2SO_4 in normal temperature conditions. The handling of pure SO_3 and oleum requires safety procedures and management in order to avoid atmospheric pollution in the case of an accidental release.

Important considerations with regard to the ancillary operations referred to above, are as follows:-

— The receipt, handling and storage of powdered raw materials should be carried out so as to minimise the emission of dust. Liquid and gaseous feeds should be carefully contained to prevent the emission of odorous fumes or gases

— Oleum and SO_3 storage and handling operations, which are often linked with H_2SO_4 production, should be installed with a means of controlling fume emissions. Venting should be directed towards acid tanks or scrubbing systems. Installations should be built by following the best engineering practice. The emissions can condense and solidify in cool areas so this must be very carefully guarded against to prevent over-pressurisation of storage tanks.

— During storage and handling of sulphuric acid, leaks may have an impact on the soil or on waters. Precautions have to be taken in order to reduce the possibility and the gravity of these leaks.

Phosphoric Acid

Processes with different raw materials are used in the manufacture of phosphoric acid. The process is known as "thermal" when the raw material is elemental phosphorus. This process has been abandoned because of the amount of energy which is needed. The processes that use phosphated minerals which are decomposed with an acid, are known as "wet processes" and they are the only economic alternative way to produce phosphoric acid.

There are three possible subgroups of wet processes depending on the acid that is used for the acidulation.

Description of Production Processes

Raw Materials for Phosphoric Acid Production

Bones used to be the principal natural source of phosphorus but phosphoric acid today is produced from phosphatic ores mined in various parts of the world. The phosphate minerals in both types of ore are of the apatite group, of which the most commonly encountered variants are:-

– Fluorapatite $Ca_{10}(PO_4)_6(F,OH)_2$

– Francolite $Ca_{10}(PO_4)_{6-x}(CO_3)_x(F,OH)_{2+x}$

Fluorapatite predominates in igneous phosphate rocks and francolite predominates in sedimentary phosphate rocks.

The most easily mined phosphate deposits are found in the great sedimentary basins. These sedimentary deposits are generally associated with matter derived from living creatures and thus contain organic compounds. These phosphates are interposed with sedimentary strata of the waste materials interpenetrated by gangue minerals and thus sedimentary phosphate ores have differing compositions within the same source. Most phosphate ores have to be concentrated or beneficiated before they can be used or sold on the international phosphate market. Different techniques may be used at the beneficiation stage, to treat the same or for removal of the gangue and associated impurities. This gives rise to further variations in the finished ore concentrate product. Phosphoric acid technology, having to rely on raw materials of great variety, has to readapt itself constantly.

Principles of the Process

The basic chemistry of the wet process is exceedingly simple. The tricalcium phosphate in the phosphate rock is converted by reaction with concentrated sulphuric acid into phosphoric acid and the insoluble salt calcium sulphate.

$$Ca_3(PO_4)_2 + 3H_2SO_4 \; 2H_3PO_4 + 3CaSO_4$$

The insoluble calcium sulphate is then separated from the phosphoric acid, most usually by filtration.

The reaction between phosphate rock and sulphuric acid is self-limiting because an insoluble layer of calcium sulphate forms on the surface of the particles of the rock. This problem is kept to a minimum by initially keeping the rock in contact with recirculated phosphoric acid to convert it as far as possible to the soluble monocalcium phosphate and then precipitating calcium sulphate with sulphuric acid.

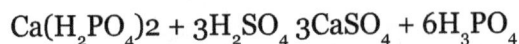

$$Ca_3(PO_4)_2 + 4H_3PO_4 \; 3Ca(H_2PO_4)_2$$

$$Ca(H_2PO_4)2 + 3H_2SO_4 \; 3CaSO_4 + 6H_3PO_4$$

Calcium sulphate exists in a number of different crystal forms depending particularly on the prevailing conditions of temperature, P_2O_5 concentration and free sulphate content The operating conditions are generally selected so that the calcium sulphate will be precipitated in either the di-hydrate or the hemihydrate form, 26-32% P_2O_5 at 70-80°C for dihydrate precipitation and 40-52% P_2O_5 at 90-110°C for hemihydrate precipitation. There are many impurities in phosphate rock, the amounts and proportions of which are very variable. The ill effects of some are manifested in the reaction system, whereas others are predominantly seen in the filtration or in the properties of the product acid. Fluorine is present in most phosphate rocks to the extent of 2-4% by weight. This element is liberated during acidulation, initially as hydrogen fluoride but in the presence of silica this readily reacts to form fluosilicic acid, H_2SiF_6. Other components such as magnesium and aluminium can also react with HF to form compounds ($MgSiF_6$ and H_3AlF_6).

A proportion of the fluorine is evolved as vapour, depending on the reaction conditions and the rest remains in the acidic medium. Some of this remainder may be precipitated by interaction with other impurities sufficiently quickly to be removed in the filter and a further proportion may subsequently contribute to sludge formation in the product acid.

More volatile fluorine compounds will appear in the vapours exhausted from the evaporators when the acid from the filter is concentrated.

Emphasis must also be placed on another group of impurities such as arsenic, cadmium, copper, lead, nickel, zinc and mercury, which are present in most phosphate rocks and which may pass into the acid during acidulation. Phosphate rocks contain naturally- occurring uranium and the radioactive components of the uranium decay series are associated with the phosphate material. The uranium goes into the product acid solution and any radium is co-precipitated with the phos-phogypsum. The amount of uranium is practically nil in some phosphate rocks. Impurities such as iron, aluminium, sodium, potassium, chlorine, etc have some influence during the production of phosphoric acid and on the quality of the acid produced.

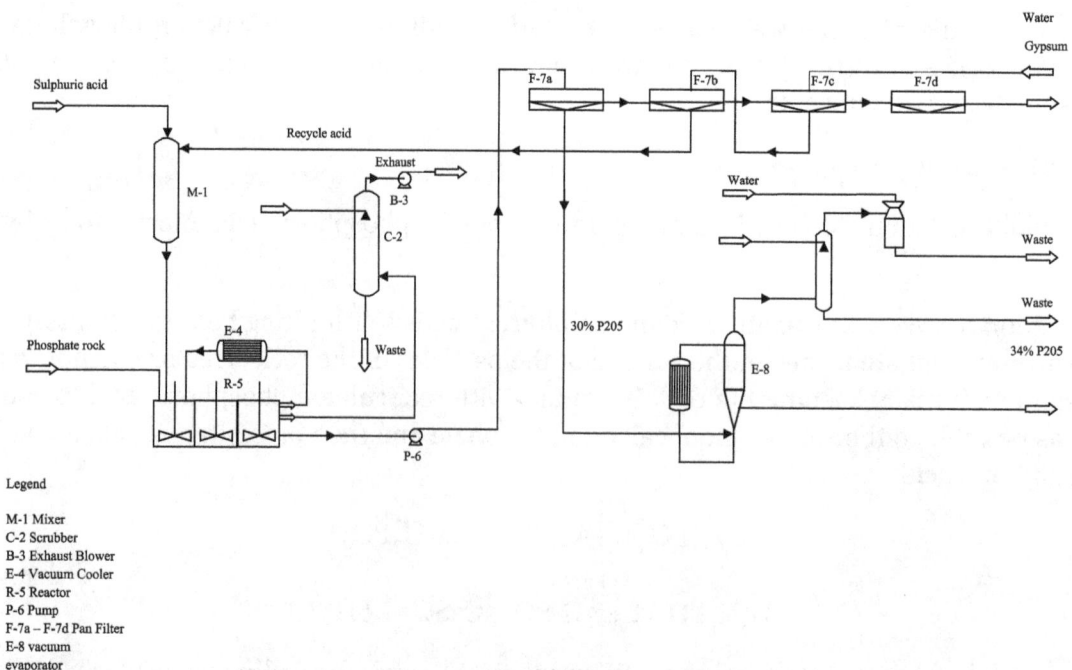

Legend

M-1 Mixer
C-2 Scrubber
B-3 Exhaust Blower
E-4 Vacuum Cooler
R-5 Reactor
P-6 Pump
F-7a – F-7d Pan Filter
E-8 vacuum
evaporator

Typical Wet-process Phosphoric Acid Plant

Production Processes of Phosphoric Acid

Five process routes are discussed and these represent the principal process routes which are available.

Dihydrate Process

This is the most diffused process and the advantages of dihydrate systems are:-

– There is no phosphate rock quality limitation

– On-line time is high

– Operating temperatures are low

– Start-up and shut-down are easy

– Wet rock can be used (saving drying costs)

The disadvantages are:-

– Relatively weak product acid (26-32% P_2O_5)

– High energy consumption in the acid concentration stage

– 4-6% P_2O_5 losses, most of them co-crystallised with the calcium sulphate

Diagram 1: Dihydrate Process

The dihydrate process comprises four stages: grinding, reaction, filtration and concentration.

Grinding

Some grades of commercial rock do not need grinding, their particle size distribution being acceptable for a dihydrate reaction section (60-70% less then 150μm). Most other phosphate rocks need particle size reduction, generally by ball or rod mills. Both mills can operate with wet or dry rock.

Reaction

The tricalcium phosphate is converted by reaction with concentrated sulphuric acid into phosphoric acid and insoluble calcium sulphate. The reactor maintains an agitated reaction volume in circulation. The reaction system consists of a series of separate agitated reactors but in the interests of economy of materials and space, the multi-vessel reaction system is replaced by a single tank in some processes. Some of these single tanks may be divided into compartments which are virtually separate reactors. The operating conditions for dihydrate precipitation are 26-32% P_2O_5 and 70-80°C. This temperature is controlled by passing the slurry through a flash cooler, which also de-gasses the slurry and makes it easier to pump. The temperature can also be controlled by using an air circulating cooler.

Filtration

This stage separates the phosphoric acid from the calcium sulphate dihydrate. Five tones of gypsum are generated for every tonne (P_2O_5) of product acid produced. The filter medium must move lation piping. A fluosilicic acid scrubber is usually included in the forced circulation evaporator system. All the evaporators in this service are generally of the single-effect design because of the corrosive nature of phosphoric acid and the very high boiling point elevation. The heat exchangers are fabricated from graphite or stainless steel with the rest of the equipment made from rubber-lined steel. All equipment designs will be made using the best practices of engineering available. More than one evaporator may be used, with the acid passing in sequence through each, depending on the degree of concentration required.

Hemihydrate (HH) Process

Operating conditions are selected in this process so that the calcium sulphate is precipitated in the hemihydrate form. It is possible to produce 40-52% P_2O_5 acid directly, with consequent valuable savings in energy requirements. The stages are similar to those of the dihydrate process but grinding may be unnecessary.

The main advantages of this process, apart from the reduction or elimination of evaporation heat requirement, are:-

Capital savings. Purer acid.

Acid from the HH process tends to contain substantially less free sulphate and suspended solids and lower levels of aluminium and fluorine than evaporated dihydrate process acid of the same strength.

Lower rock grinding requirements.

A satisfactory rate of reaction can be achieved from much coarser rock than in the dehydrate process, because of the more severe reaction conditions in the HH process.

The disadvantages of HH systems are:-

Filtration rate.

Hemihydrate crystals tend to be small and less well formed than dihydrate crystals and thus hemihydrate slurries tend to be more difficult to filter than dihydrate slurries unless crystal habit modifiers are used to suppress excessive nucleation. With a good HH process however, there is no need to use crystal habit modifiers. There are examples of phosphate rocks that produce hemihydrate crystals achieving higher filtration rates than obtained with dihydrate crystals.

Phosphate losses.

Water balance considerations restrict the amount of wash water that can be used. At the same time, the amounts of both soluble and insoluble P_2O_5 remaining in the filter cake are greater because of the higher P_2O_5 concentration of the slurry being filtered. Nevertheless the simplicity of the HH plant and the absence of silicofluoride and chucrovite scaling in the HH filter, may compensate for the higher insoluble P_2O_5 loss via HH cake.

Scaling.

Hemihydrate is not a stable form of calcium sulphate and there is a tendency for it to revert to gypsum even before the acid has been filtered off. The conditions are even more in favour of rehydration during washing. In a good HH plant there is no conversion in the reactor battery. A small quantity of anti-scale agent may be required in a single-stage HH plant filter to avoid scaling.

Filter cake impurity.

The cake is more acidic than gypsum filter cake because of the extra P_2O_5 losses and it also contains more fluorine and cadmium.

Corrosion.

The demands on susceptible items of equipment, particularly agitators and slurry pumps, are increased because of the higher temperature (100°C) and acid concentration (40-50% P_2O_5) compared to a dihydrate plant.

Recrystallisation Processes

The best P_2O_5 recovery efficiencies that generally can be expected in the single-stage dihydrate and hemihydrate processes considered so far, are 94-96% and 90-94% respectively. The P_2O_5 losses are retained in the filter cake and this can create problems with disposal or use of the by-product gypsum. Some of this loss of P_2O_5 passes into solution and can be recovered when the calcium sulphate is finally separated, if the calcium sulphate is made to recrystallise to its other hydrate. This not only raises the overall efficiency of the process but also gives a much cleaner calcium sulphate.

Several processes have been developed but there are only three basic routes:-

– Acidulate under hemihydrate conditions; recrystallise to dihydrate without intermediate hemihydrate separation; separate product. (Hemihydrate recrysallisation (HRC) process)

– Acidulate under hemihydrate conditions; separate product; recrystallise hemihydrate to dihydrate; filter and return liquors to process. (Hemi-dihydrate (HDH) process)

– Acidulate under dihydrate conditions; separate product; recrystallise hemihydrate; filter and return liquors to process (Dihydrate-Hemihydrate (DH/HH) process)

HRC Process

The flow diagram of this process resembles that of the multiple reactor dihydrate process with the exception that the attack reactor operates under hemihydrate conditions, while succeeding reactors operate under conditions favouring the rehydration of hemihydrate to gypsum. This is encouraged by seed dihydrate crystals recycled in the slurry from the filter feed. The product acid is no more concentrated than that obtained from dihydrate but the gypsum is much purer.

HDH Process

It is possible to obtain $40-52\%$ P_2O_5 acid directly, by acidulating under hemihydrate conditions and separating the hemihydrate before recrystallising, in this process. The additional filter and the other equipment required, add to the capital cost of the plant but enable savings to be made on evaporation equipment.

DH/HH Process

In this process, although the reaction runs under dihydrate conditions, it is not desirable to effect a very high degree of P_2O_5 recovery during the separation of the acid from the dihydrate. The succeeding dehydration stage requires around $20-30\%$ P_2O_5 and 10-20% sulphuric acid. The strength of the product acid is $32-35\%$ P_2O_5.

Repulping Process

A further optimisation of the HRC process can be obtained by re-slurrying and washing the gypsum, followed by a second filtration step in the so called "Repulping Process". Most of the free acid which is not removed in the first filtration step, can be removed in this process and the efficiency can be improved by up to 1% (depending on the amount of free acid). The gypsum from the first filter is re-slurried in a tank and then pumped to a second filter where the gypsum is de-watered. The gypsum is then washed with the fresh water coming into the plant. The liquid obtained from the second filter is used on the first filter to wash the gypsum. The repulp process is in fact an additional step in the counter-current washing of the gypsum using the water that enters the plant.

Monocalcium Phosphate

Monocalcium phosphate is an inorganic compound with the chemical formula $Ca(H_2PO_4)_2$ ("ACMP" or "CMP-A" for anhydrous monocalcium phosphate). It is commonly found as the monohydrate ('"MCP" or "MCP-M"), $Ca(H_2PO_4)_2 \cdot H_2O$ (CAS# 10031-30-8). Both salts are colourless solids. They are used mainly as superphosphate fertilizers and are also popular leavening agents.

Preparation

Material of relatively high purity, as required for baking, is produced by treating calcium hydroxide with phosphoric acid:

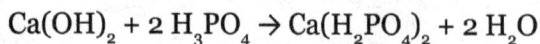

$$Ca(OH)_2 + 2\ H_3PO_4 \rightarrow Ca(H_2PO_4)_2 + 2\ H_2O$$

Samples of $Ca(H_2PO_4)_2$ tends to convert to dicalcium phosphate:

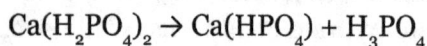

$$Ca(H_2PO_4)_2 \rightarrow Ca(HPO_4) + H_3PO_4$$

Applications

Use in Fertilizers

Superphosphate fertilizers are produced by treatment of "phosphate rock" with acids. Using phosphoric acid, fluorapatite is converted to $Ca(H_2PO_4)_2$:

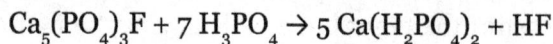

$$Ca_5(PO_4)_3F + 7\ H_3PO_4 \rightarrow 5\ Ca(H_2PO_4)_2 + HF$$

This solid is called triple superphosphate. Several million tons are produced annually for use as fertilizers. Residual HF typically reacts with silicate minerals co-mingled with the phosphate ores to produce hydrofluorosilicic acid (H_2SiF_6). The majority of the hexafluorosilicic acid is converted to aluminium fluoride and cryolite for the processing of aluminium. These materials are central to the conversion of aluminium ore into aluminium metal.

When sulfuric acid is used, the product contains phosphogypsum ($CaSO_4 \cdot 2H_2O$) and is called single superphosphate.

Use as Leavening Agent

Calcium dihydrogen phosphate is used in the food industry as a leavening agent, i.e., to cause baked goods to rise. Because it is acidic, when combined with an alkali ingredient, commonly sodium bicarbonate (baking soda) or potassium bicarbonate, it reacts to produce carbon dioxide and a salt. Outward pressure of the carbon dioxide gas causes the rising effect. When combined in a ready-made baking powder, the acid and alkali ingredients are included in the right proportions such that they will exactly neutralize each other and not significantly affect the overall pH of the product. AMCP and MCP are fast acting, releasing most carbon dioxide within minutes of mixing. It is popularly used in pancake mixes. In double acting baking powders, MCP is often combined with the slow acting acid sodium acid pyrophosphate (SAPP).

Nitrophosphate Fertilizers

"Nitrophosphate" is the generally accepted term for any fertilizer that is produced by a process involving treatment of phosphate rock with nitric acid.

Fundamentals of Nitrophosphates

The basic (and simplified) acidulation reaction can be represented by the following reaction equations:

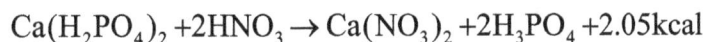

$$Ca_3(PO_4)_2 + 4HNO_3 \rightarrow 2Ca(NO_3)_2 + Ca(H_2PO_4)_2 + 20.2\text{kcal}$$

$$Ca(H_2PO_4)_2 + 2HNO_3 \rightarrow Ca(NO_3)_2 + 2H_3PO_4 + 2.05\text{kcal}$$

Overall

$$Ca_3(PO_4)_2 + 6HNO_3 \rightarrow 3Ca(NO_3)_2 + 2H_3PO_4 + 22.25\text{kcal}$$

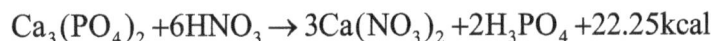

After separation of the insoluble material, phosphoric acid is neutralized with ammonia to produce a fertilizer. If the calcium nitrate is left in the solution, it reverts to dicalcium phosphate ammonium following the reaction:

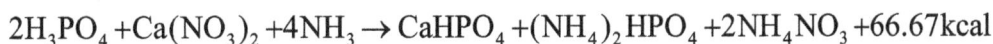

$$2H_3PO_4 + Ca(NO_3)_2 + 4NH_3 \rightarrow CaHPO_4 + (NH_4)_2HPO_4 + 2NH_4NO_3 + 66.67\text{kcal}$$

Calcium is non-nutrient in terms of N, P, and K and is therefore seen as a diluent. For these reasons alone, it is desirable to remove calcium from the solution. But there is another, more important reason. If the calcium nitrate is left in the solution, when it is neutralized the N:P ratio will be fixed. To produce grades with lower N:P ratios, it is necessary to remove calcium nitrate, whereas if higher nitrogen grades are required, the removed calcium nitrate can be converted to ammonium nitrate and returned to the solution. Therefore, most nitrophosphate processes include some means of removing calcium nitrate from the solution.

Nitrophosphate Process

Selection of Phosphate Rock

In general, reactivity of the phosphate rock is no problem; even igneous apatites dissolve readily in nitric acid. The rock need not be finely ground; rock finer than 1 mm is satisfactory, and some operators even accept particles up to 4 mm. In general, the rock need only be fine enough to prevent rapid settling in stirred reaction vessels. High-silica rock can be used if the equipment is designed for that purpose. Most nitrophosphate plants include a silica removal step. Coarse silica particles can be very abrasive to pumps and piping, and this fact should be considered in plant design.

It is desirable that the $CaO:P_2O_5$ ratio in the rock should be as low as economically feasible to minimize the amount of calcium that must be removed or offset(in mixed-acid processes). While additional calcium requires more nitric acid, it does not necessarily involve a direct economic penalty because the nitrate is subsequently converted to ammonium nitrate either in the nitrophosphate product or in a coproduct.

Carbonates is phosphate rock cause foaming, which is usually dealt with by using mechanical foam breakers. However, foaming can be a difficult problem with some rocks.

Organic matter is undesirable in nitric phosphate processes; it reacts with nitric acid with emission of nitrogen as NO_2 or other nitrogen oxides.

Iron and aluminum oxides present no special problem within the range of occurrence in commercial phosphate rocks; these oxides usually are dissolved in nitric acid and reprecipitated during ammoniation as citrate-soluble phosphates. TVA tested "leached- zone" Florida phosphate containing a high percentage of aluminum phosphate minerals in a special nitrophosphate process.

Process Alternatives

The two commercially important nitrophosphate processes differ in the way they solve the problem of phosphate water solubility caused by the presence of calcium nitrate in the slurry resulting from the reaction of phosphate rock with nitric acid. The first process, historically, is the Odda process wherein calcium nitrate is precipitated and separated. The "mixed acid " process does not separate the calcium nitrate; the phosphate water solubility is increased by adding phosphoric acid to decrease the $CaO:P_2O_5$ ratio.

Odda Process

The nitrophosphate process (also known as the Odda process) was a method for the industrial production of nitrogen fertilizers invented by Erling Johnson in the municipality of Odda, Norway around 1927.

The process involves acidifying phosphate rock with nitric acid to produce a mixture of phosphoric acid and calcium nitrate.

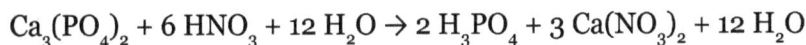

$$Ca_3(PO_4)_2 + 6\ HNO_3 + 12\ H_2O \rightarrow 2\ H_3PO_4 + 3\ Ca(NO_3)_2 + 12\ H_2O$$

The mixture is cooled to below 0°C, where the calcium nitrate crystallizes and can be separated from the phosphoric acid.

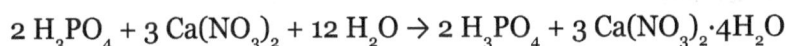

$$2\ H_3PO_4 + 3\ Ca(NO_3)_2 + 12\ H_2O \rightarrow 2\ H_3PO_4 + 3\ Ca(NO_3)_2 \cdot 4H_2O$$

The resulting calcium nitrate produces nitrogen fertilizer. The filtrate is composed mainly of phosphoric acid with some nitric acid and traces of calcium nitrate, and this is neutralized with ammonia to produce a compound fertilizer.

$$Ca(NO_3)_2 + 4\ H_3PO_4 + 8\ NH_3 \rightarrow CaHPO_4 + 2\ NH_4NO_3 + 3(NH_4)_2HPO_4$$

If potassium chloride or potassium sulfate is added, the result will be NPK fertilizer. The process was an innovation for requiring neither the expensive sulfuric acid nor producing gypsum waste.

The calcium nitrate mentioned before, can as said be worked up as calcium nitrate fertilizer but often it is converted into ammonium nitrate and calcium carbonate using carbon dioxide and ammonia.

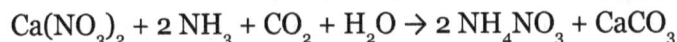

$$Ca(NO_3)_2 + 2\ NH_3 + CO_2 + H_2O \rightarrow 2\ NH_4NO_3 + CaCO_3$$

Both products can be worked up together as straight nitrogen fertilizer.

Although Johnson created the process while working for the Odda Smelteverk, his company never employed it. Instead, it licensed the process to Norsk Hydro, BASF, Hoechst, and DSM. Each of these companies used the process, introduced variations, and licensed it to other companies. Today, only Yara (Norsk Hydro), BASF, Borealis Agrolinz Melamine GmbH, and GNFC still use the Odda process. Due to the alterations of the process by the various companies who employed it, the process is now generally referred to as the nitrophosphate process.

Due to the byproduct ammonium nitrate which has lower value, the production of ammonium nitrophosphate fertilizer is not economical compared to diammonium phosphate which is produced from cheaper sulfuric acid or gypsum.

The final slurry after ammoniation and evaporation is formed into granules, with or without addition of potash salts, by a variety of methods including:

- Granulation in a pugmill or blunger.
- Ammoniation and granulation in a rotary drum.
- Granulation and drying in a Spherodizer®.
- Granulation and drying in a spouted bed system

Prilling of a melt. Calcium Nitrate Conversion and Calcium Ammonium Nitrate (CAN) Unit

The CN-conversion and CAN units contain the following sections:

- CN Conversion Unit
    - Ammonium carbonate preparation section
    - CN-conversion section
    - Lime separation section
- CAN Unit
    - AN concentration section
    - Additives preparation section
    - Mixing, granulation (or prilling), and drying section
    - Cooling, screening, and coating section
    - Off-gas Treatment Section

CN-conversion section begins with the preparation of ammonium carbonate solution in a dilute ammonium nitrate solution in a packed carbonation tower with external circulation. Heat exchangers are used to control the temperature gradient over the tower. Raw materials are carbon dioxide and gaseous ammonia evaporated in the refrigeration section. The required ammonium carbonate solution is taken from the circulation loop and sent to the conversion reactor with the corresponding the quantity of CN solution. A lime settler and a vacuum filter separate the reaction products, dilute ammonium nitrate solution, and calcium carbonate 9lime). The AN- solution obtained here requires a second filtration step on precoat is established with the product lime itself. After this second filtration step the remaining ammonium nitrate is stored in the AN-solution tank; from there it is pumped to the AN evaporation section and to the CNTH-filter to dissolve fresh calcium nitrate tetrahydrate crystals.

The CN-conversion section has its own packed off-gas scrubber with external circulation. The scrubber liquid is an acidified dilute AN-solution, which is returned to the process.

The AN-evaporation section consists of several falling film evaporators in series. The first part of this section produces a concentrated AN-solution of 93-94 wt% AN. Depending on the capacity and cost of energy, this step may consist of double- or triple-effect evaporation. Each falling film evaporator set consists of a preheater, heater/ evaporator, vapor separator and transfer pump. A two-stage water ring vacuum pump provides the necessary vacuum, and a steam saturator, operating at 0.9 MPa absolute (175°C condensing temperature), provides the necessary energy for the concentration unit.

The ammonium nitrate solution at 93-94 wt % is stored and pumped to the second part of the evaporation unit and/or to the NP-unit. This second part is composed of one falling film evaporator with the separator operating at 0.3 bar (absolute) and concentrating up to the required concentration of about 98 wt %. Steam is removed from a separate steam saturator at 9 bar(absolute). The vacuum system of this section can be combined with that of the first part.

CAN-slurry is produced in a mixing section, which consists of dissolving vessels for additives and feeding equipment for gypsum, which is added to the lime. Dissolved additives, lime/gypsum mixture and concentrated AN-melt are delivered to a mixing vessel with an agitator. From there it flows into a second vessel and is then circulated back to the mixing vessel. CAN slurry for granulation is taken from this circulation loop.

The CAN granulation section uses a granulation drum with internals to produce a screen of falling materials, on which the concentrated CAN slurry is sprayed. The granules that are produced fall by gravity into a counter currently operated drying drum. The drying drum is operated autothermally, which means that no external heat source is required to preheat the drying air. Off-gas from the dryer and granulator is first dedusted in cyclones, and then sent to the same gas scrubber used for scrubbing the off-gas from the CN-conversion section.

Dry granules are screened over double-deck screens. Fines, crushed oversize, and part of the on size are returned to the granulator. A two-stage fluidized bed cooler, of which the second stage is operated with chilled and conditioned air, cools the product to the desired final temperature. The waste air from the fluidized bed cooler must be treated to remove dust.

Coating takes place in a rotating coating drum to improve product characteristics during storage. From the coating drum, product is sent to storage.

The BASF granulation process for CAN can use either "AN-wet" lime – lime from the lime filter, which is washed with AN solution, or dry lime, which is water- washed lime. For either alternative the residual water content of the AN-melt must be adjusted.

The advantage of using "AN-wet" lime is that a considerable amount of energy and capital is saved. A saving of energy occurs because no dilution of AN-solution with wash-water occurs and no lime drying takes place. Investment is reduced because lime drying equipment is not required.

In prilling CAN, the AN solution is premixed with dry lime immediately before prilling. A rotating, perforated bucket is the preferred type of drop-forming apparatus. Prill towers for both CAN and AN commonly are 45-56 m high although shorter ones are also used. For high-density prilling using 99.7% solution, prilling towers that are 15-30 m tall may be used. Cooling may be conducted in a rotary coler or in a fluidized bed either in the bottom of the prill tower or in a separate unit.

Mixed-Acid Process for Preparation of Nitrophosphates

Phosphate Rock Digestion and Ammoniation

The first step of the process is the digestion of phosphate rock with nitric acid, which results in a solution of phosphoric acid and calcium nitrate (first reactor).

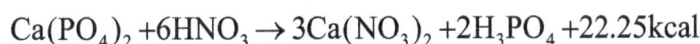

$$Ca(PO_4)_2 + 6HNO_3 \rightarrow 3Ca(NO_3)_2 + 2H_3PO_4 + 22.25kcal$$

Depending on the type of phosphate rock, formation of acid gases containing, for example, nitrogen oxides and compounds of fluorine takes place during the digestion. In the second reactor the overflow from the first reactor is ammoniated, and scrubber liquor is added.

The acid is slurry is ammoniated with gaseous ammonia along the following reactions.

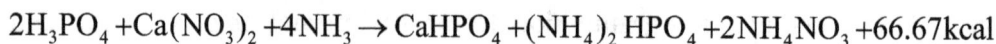

$$HNO_3 + NH_3 \rightarrow NH_4NO_3 + 22.27 \text{kcal}$$

$$H_3PO_4 + 2NH_3 \rightarrow (NH_4)_2 HPO_4 + 33.52 \text{kcal}$$

$$2H_3PO_4 + Ca(NO_3)_2 + 4NH_3 \rightarrow CaHPO_4 + (NH_4)_2 HPO_4 + 2NH_4NO_3 + 66.67 \text{kcal}$$

The third reactor provides for final ammoniation of overflow from the second reactor.

Phosphoric and sulfuric acids are added and the KCl dissolved (when NPK is produced).

If sulfuric acid is introduced, the sulfate ions induce the precipitation of calcium sulfate following the reaction:

$$Ca(NO_3)_2 + H_2SO_4 \rightarrow 2HNO_3 + CaSO_4 + 9.37 \text{kcal}$$

This calcium sulfate can be removed by filtration, but generally it is left in the slurry where it may be considered as a diluent that supplies two secondary nutrients: calcium and sulfur.

Any micronutrients to be incorporated in the formulation are added at this point. A buffer tank is used as a liquor feed tank for the Spherodizer® and as a receptacle for recovering fertilizer dust from the bag filters and cyclones.

The water solubility of P_2O_5 in the product can be adjusted by varying the phosphoric acid/phosphate rock ratio in the formulation. A wide range of grades may be produced; these grades include: for minimum nitrogen content, 8-24-14 grade; and for maximum nitrogen content, 30-10-0 grade. Standard grades such as 17-17-17 and 23-23-0 may be produced.

After neutralization other components, e.g., ammonium phosphates, superphosphates, ammonium sulfate and compounds containing potassium and magnesium, may be added. Most of these materials may also be added before or during neutralization, but if the raw material contains chloride, the pH-value of the slurry should be 5-6 to avoid development of hydrogen chloride.

The design of the reactor battery can vary from a few large reactors to many small reactors, but at present the three-reactor system is considered optimal. A common feature for all the designs is that the row of reactors ends with a buffer tank. Depending on the type of raw material, the amount of gas scrubber liquid to be recycled, and the degree of ammoniation, the water content of the slurry in the buffer tank can vary between 5% and 30% and the temperature from 100°C to 140°C.

Further Processing

Granulation and Drying: The slurry in the buffer tank typically contains about 10% of water and the temperature is about 140°C. Granulation and drying occur simultaneously in a combined granulator-dryer called a Spherodizer®.

Screening, Crushing, and Recycle Handling: Product sized 2-4 mm is separated from the granulator discharge by screens. The oversize material is crushed and recycled with undersize to the granulator.

Nutrient recovery From Gas Streams: A multistage scrubbing system captures fumes from the reactors and any dust that escapes the cyclones. The nutrient content of the scrubber liquor increases as it moves counter currently through the reactor gas scrubbing system to the granulator exhaust gas scrubber.

Advantages and Disadvantages of the Nitrophosphate Route

The relative economics of fertilizer production by the nitrophosphate and sulfuric routes have been a matter of discussion for many years, especially at times of sulfur shortage when prices tend to be high. The balance swung sharply away from nitrophosphates and in favor of the sulfuric route for most of the 1960s and 1970s. The reason for this change was first, because sulfur prices declined and then remained stable at a low level for several years; the second reason was because of the sharp rise in-energy-and-ammonia feedstock prices (which directly affect nitric acid production costs) from about 1974 onward. Interest revived considerably when a sulfur shortage caused prices to increase very sharply. Although the acute phase of the shortage proved to be short lived, sulfur prices have settled to a level that is fairly high in comparison with their in 1970s levels; on the other hand, energy and feedstock prices have declined considerably from their peak. In the mid-1990s, sulfur prices declined, but energy prices remained stable, thus creating a more balanced competition. Under these conditions, several other factors besides mere process economics can influence the ultimate choice of the nitrophosphate route; just recently it has been evident that this has been happening in some cases.

Prominent among the other factors are the comparative logistics and security of sulfur and ammonia feed-stock supplies and the difficulty and cost of disposing the phosphogypsum byproduct. The former was important in the choice of the process for large fertilizer complexes in central Europe and the Indian subcontinent in the 1960s and 1970s. The latter is assuming greater importance today in many areas as a result of environmental regulations.

As a result of developments in recent years, the nitrophosphate process of today is more economically and environmentally competitive. It claims advantages such as raw material flexibility and efficiency, adaption to increasingly stricter environmental requirements, and avoidance of byproduct disposal problems.

However, the process has its limitations in that it combines phosphate and nitrogen processing. The advantage of being sulfur independent is counterbalanced by the need for carbon dioxide, which is normally obtained from an ammonia plant.

Other Phosphate Fertilizers

Enriched Superphosphate

"Enriched" superphosphate is essentially a mixture of SSP and TSP, usually made by acidulation of phosphate rock with a mixtur of sulfuric and phosphoric acids. Theoretacally, any grade between SSP and TSP can be produced, but the usual range is 25% - 35% P_2O_5. Processes and equipment are about the same as for SSP[4].

Enriched superphosphate may be a useful product for application in sulfur-deficient areas where SSP would supply more sulfur than necessary. One advantage is that mixed acid of the proper concentration can be obtained by mixing concentrated sulfuric acid(93% or 98% H_2SO_4) with dilute phosphoric acid(30% P_2O_5), thereby avoiding the need for concentrating the latter.

Ground Phosphate rock is effective only on acid soils (pH 6 or less). This statement applies to apatitic rocks not to calcined aluminum phosphate ores that are effective on neutral or calcareous soils.

Phosphate rocks are also more effective in warmer climates, in moist soils, and on crops that have fairly long-term growing patterns. They are less effective for short-season crops grown under cool soil temperatures, particularly in the first year of application.

It is generally agreed that the rock should be finely ground and well mixed with the soil. However, there is some difference of opinion as to the usefulness of very fine grinding. The majority opinion seems to be that there is little to be gained by grinding finer than about 90% through 100-mesh (0.147 mm), although there are some who claim that very fine grinding such as 80% through 320-mesh (0.043 mm) is worthwhile.

Ground rock that has been granulated after grinding has given rather poor first-crop results even when the granules disintegrate in the soil, presumably because granulation reduces the area of contact with the soil. However, in most of the tests, the granules have been in the range of 1-4 mm.

While it is generally agreed that reactivity is important, there is some disagreement as to how important it is and how to measure it. Chemical methods for evaluating reactivity are discussed later. The importance of reactivity is greatest for the first crop or season; the long-term or residual effectiveness does not seem to be closely related to reactivity. Rocks of relatively low reactivity have shown good long-term effectiveness.

Some investigators have attributed the response to phosphate rock to annul rainfall; better results were obtained on well-watered soils.

Ground phosphate rock has been advocated and used for reclaiming low-phosphorus soils of abandoned farms or of new (previously uncultivated) land of low native phosphorus fertility.

Calcination is carried out in a fluidized bed. Experiments showed that maximum citrate solubility was obtained in the range of 400° − 600°C. The maximum solubility in alkaline ammonium citrate solution was about 70%; the solubility in 2% citric acid solution is much lower, about 20% - 30% . The product is used for direct application after being ground to pass a 100-mesh (0.15mm) screen.

Basic Slag

Basic slag, also called Thomas slag, is a byproduct of the steel industry. Iron made from high-phosphorus ore is converted to steel in a Thomas converter by oxidation in contact with a basic (high CaO) slag. The usual range of P_2O_5 content in slag that is used for phosphate fertilization is 10% - 20%. Sometimes phosphate rock is deliberately added to the blast furnace charge to increase the phosphorus content of the iron and thereby increase the P_2O_5 content of the slag.

Basic open-hearth slag also may contain P_2O_5 up to 10% - 20%, and it is used in agriculture in some countries, both for liming and phosphorus supply.

The P_2O_5 in basic slag is mainly present as calcium silicophosphates – silicocarnotite ($5CaO \cdot P_2O_5 \cdot SiO_2$) and nagelschmitite ($7CaO \cdot P_2O_5 \cdot SiO_2$). Small amounts of fluorspar (CaF_2) may be added to slags to decrease their viscosity during the steel-refining process. Such slags contain fluorapatite and are likely to be less suitable for fertilizer use.

Potassium Phosphates

Potassium Phosphates are excellent fertilizers, and their very high analysis is an advantage that has stimulated much research in an effort to find an echonomical production process. However, no process has been developed that is economical enough to result in widespread production; therefore, present use is limited to special purposes for which the high cost can be justified.

At present, most of the potassium phosphates used in fertilizers are produced from potassium hydroxide or carbonate and phosphoric acid and are used in liquids for foliar application or other specialty uses.

Some of the alternative salts of potassium phosphates are given in table

Phosphate Potassium Salts		
Compound	Formula	Grade
Monopotassium phosphate	KH_2PO_4	0-52-35
Dipotassium phosphate	K_2HPO_4	0-40-54
Tetra potassium pyrophosphate	$K_4P_2O_7$	0-43-57
Potassium metaphosphate	KPO_3	0-60-40

In addition, a potassium polyphosphate solution of 0-26-27 grade has been produced from superphosphoric acid and potassium hydroxide; it contains a mixture of ortho, pyro, and higher polyphosphates.

TVA, SAI, and others have produced potassium metaphosphate in pilot plants by high- temperature reaction of KCl and phosphoric acid. The pure material, KPO , has a grade of about 0-60-40 and, thus, a 100% nutrient content (on an oxide basis).

Bone Meal

Bone meal is a mixture of crushed and coarsely ground bones that is used as an organic fertilizer for plants and formerly in animal feed. As a slow-release fertilizer, bone meal is primarily used as a source of phosphorus.

As a fertilizer, the N-P-K ratio of bone meal is generally 4-12-0, though some steamed bone meals have N-P-Ks of 1-13-0. Bone meal is also an excellent organic source of calcium. Organic fertilizers usually require the use of microbes/bacteria in the soil in order to make the nutrients in the fertilizer bio-available. That can result in irregular release of phosphorus/calcium. In sterile potting soil, there may be no microbes to release the nutrients.

Finely ground bone meal may provide quicker release than coarsely ground. Phosphates do not easily pass through soil. So mixing the bone meal with the soil or putting it in the planting hole can help.

Fused Calcium Magnesium Phosphate

If a mixture of phosphate rock and olivine or serpentine (magnesium silicate) is fused in an electric furnace . The molten product is quenched with water and used in a finely divided state as a fertilizer. The product, a calcium magnesium phosphate (CMP) glass, contains about 20% P_2O_5 and 15% MgO. Over 90% of the product is soluble in citric acid.

The theoretical compositions of some magnesium-containing minerals that can be used to produce CMP are:

Olivine	$(Mg, Fe)_2\ SiO_4$
Serpentine	$Mg_3\ H_4SiO_9$
Garnierite	$(Mg, Ni)\ H_2SiO_4$
Magnesite	$MgCO_3$

The minerals are variable in compositions; iron, nickel, and sometimes manganese may substitute for magnesium. Magnesium oxide, obtained by calcining magnesite or extracting it from sea water, can be used in the process; in this case, silica must be added in sufficient quantity to result in 20% - 30% SiO_2 in the product.

Rhenania Phosphate

Rhenania Phosphate is another thermally produced phosphate fertilizer. It is made by calcining a mixture of phosphate rock, sodium carbonate, and silica in a rotary kiln at 1250°C . Enough sodium carbonate is used to form the compound $CaNaPO_4$ and enough silica to form Ca_2SiO_4 with the excess calcium. Typical charge proportions are one part of sodium carbonate to three parts of phosphate rock and enough silica to raise the SiO_2 content of the product to about 10%. The product contains 28% − 30% P_2O_5, which is nearly all soluble in neutral or alkaline ammonium citrate solution even though much of the flurine remains in the product. It is applied to the soil in pulverized form or granulated in small granules with potash salts. Some grades are produced containing magnesium or boron, which are added during granulation as kieserite or borax, respectively.

A somewhat similar product, Roechling phosphate, uses a soda slag that is a byproduct from the steel industry. Also, the naturally occuring minerals, trona (sodium sesquicarbonate) or natron (sodium carbonate), may be used. Experimants have shown that a similar product can be made by sintering potassium carbonate with phosphate rock and silica to produce a product grade of 0-25-25. The phosphate compound in this product is presumed to be $CaKPO_4$.

The overall reaction in producing Rhenania phosphate is assumed to be:

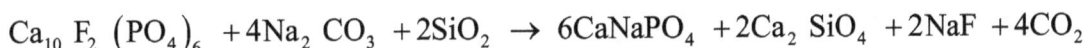

$$Ca_{10} F_2 (PO_4)_6 + 4Na_2 CO_3 + 2SiO_2 \rightarrow 6CaNaPO_4 + 2Ca_2 SiO_4 + 2NaF + 4CO_2$$

Any grade of phosphate rock can be used, but since the grade of the product is determined by the grade of the rock, a high grade is preferred.

Dicalcium Phosphate

Dicalcium Phosphate is a common constituent of nitrophosphate fertilizers and of compound fertilizers formed by ammoniation of superphosphates. The process of production consists of dissolving phosphate rock in hydrochloric acid and then precipitating dicalcium.

phosphate by stepwise addition of limestone and slaked lime. The product is recovered by filtration and washing, and the remaining solution of calcium chloride may be used or discarded.

Magnesium Phosphates

Monomagnesium, dimagnesium, and trimagnesium phosphates are known to be effective fertilizers, but there is no known commercial production of these materials for fertilizer use. No doubt small percentages of these compounds are formed in processing phosphate rock containing magnesium.

Urea Superphosphate (USP)

Urea and sulfuric acid form the following complexes:

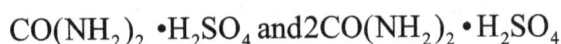

$$CO(NH_2)_2 \cdot H_2SO_4 \text{ and } 2CO(NH_2)_2 \cdot H_2SO_4$$

There are two compounds that correspond respectively to 3.6 moles of urea to 1 mole of H_2SO_4 and 1.8 mole of urea to 1 mole of acid. While the melting point of urea is 132.7°C, both have a melting point of about 10°C.

The preparation of the mixtures of urea, sulfuric acid, and water at the mole ratios, 3:6:1 and 1:8:1, is exothermic in both cases. Heat release with the first ratio is lower than with the second one and allows the preparation of the mixture under stable and reliable conditions at an equilibrium temperature of 60° C - 70° C, which is ideal to acidulate phosphate rock.

In the manufacture of USP, the reaction of acidulation may be written as follows.

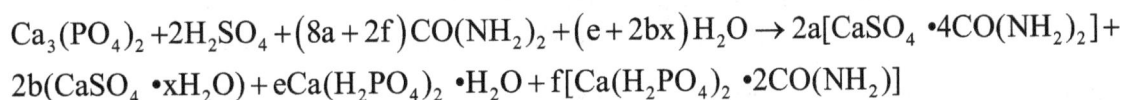

$$Ca_3(PO_4)_2 + 2H_2SO_4 + (8a+2f)CO(NH_2)_2 + (e+2bx)H_2O \rightarrow 2a[CaSO_4 \cdot 4CO(NH_2)_2] +$$
$$2b(CaSO_4 \cdot xH_2O) + eCa(H_2PO_4)_2 \cdot H_2O + f[Ca(H_2PO_4)_2 \cdot 2CO(NH_2)]$$

With a+b = 1, and e+f = 1.

It will be noticed that urea is associated with calcium sulfate rather than water of hydration. But the sulfuric acid to rock ratio has not changed.

Identification of the Reaction Products

X-ray analysis of the product showed that:

- There is no more free urea.

- There is a substantial amount of tetra urea calcium sulfate.

- P_2O_5 as monocalcium phosphate may be linked to two ureas.

Properties of USP 20-10-0 Properties	% Weight
Total nitrogen	20.9
Urea nitrogen	19.3
Ammoniacal nitrogen	1.6
Total P_2O_5	10.2
Citrate-soluble P_2O_5	9.7
Water-soluble P_2O_5	9.2
SO_3	16.1
H_2O	1.0

Storage Properties

USP can be used as produced, i.e., in powrered form, or as a granular material. In the later case, its physical properties are quite similar to those of urea-based NP and NPK grades. The product stores well. Its critical relative humidity is 65% - 70% at 200C; consequently, it is suitable for bulk storage.

Agronomic Interest

To determine whether this new fertilizer with urea in the molecular structure has agronomic properties typical traditional fertilizers, agronomic tests were done with maize and rice.

USP was tested in comparision with the same quantities of nitrogen and phosphate supplied by DAP and urea. In both cases the same input of potassium was supplied by potassium chloride (KCl) containing 60% K_2O. Tests were carried out in five different combination of clay, silt, sand, and loam soils.

An increase in yeild was found at the opotimum nitrogen input. Thus, for a yield of 11,500 kg/ha of maize, the fertilization rate can be reduced by 40 kg/ha of nitrogen when using USP.

USP of the 20-10-0 grade with 60% K_2O potassium chloride was compared with a mixture comprising ammonium sulfate, TSP, and potassium chloride. Only one type of soil was used for this comparison; a loam soil composed of clay, sand and silt. With the same fertilization rate of 140 kg/ha nitrogen, the yield was increased by 10%. The tests also showed that splitting the application of nitrogen, which was advisable in traditional fertilization, was no longer necessary with USP.

The tests have shown that USP affects maize and rice fertilization very positively. At constant yield the consumption of nitrogen can be cut by at least 10%.

Advantages of the Process

- Zero liquid effluents, near zero flurine emission.

  Because the flurine in the phosphate rock is entirely recovered in the USP, a single-stage scrubbing unit satisfies the most stringent standards while SSP requires three or four stages. Moreover, the scrubbing liquor is recycled into the preparation of the urea-sulfuric acid mixture.

- The phosphoric acid route is avoided.

  This is one of the rare processes that allow for the production of a urea-based compound fertilizer without using phosphoric acid, thus avoiding its costs and nuisances.

- Granulation plants can be retrofitted to operate the USP process.

- The USP process cost effective.

  USP is cost effective because it uses the lowest cost raw materials based on concentrated sulfuric acid, it produces a drier product that rquires no additional drying when used in powdered form or saves 40% of the drying energy when it is granulated. The production technology is simple and requires limited capital cost.

- It can be produced in SSP or TSP plants after an easy and cheap revamping.

Potash and Potassium Fertilizers

Potassium is an important crop nutrient. It improves yield, water retention, taste and disease resistance in food crops. Potassium can be applied to sugar, corn, wheat, rise, vegetables and fruits. The chapter on potash and potassium fertilizers offers an insightful focus, keeping in mind the complex subject matter.

Potash

Potash is any of various mined and manufactured salts that contain potassium in water-soluble form. The name derives from *pot ash*, which refers to plant ashes soaked in water in a pot, the primary means of manufacturing the product before the industrial era. The word *potassium* is derived from potash.

Crystals of potash with a coin for reference. The coin (a US penny) is 19 mm (0.75 in) in diameter and copper in color.

Potash is produced worldwide at amounts exceeding 30 million tonnes per year, mostly for use in fertilizers. Various types of fertilizer-potash thus constitute the single largest global industrial use of the element potassium. Potassium was first derived by electrolysis of caustic potash (a.k.a. potassium hydroxide), in 1807.

Terminology

Potash refers to potassium compounds and potassium-bearing materials, the most common being potassium chloride (KCl). The term *potash* comes from the Middle Dutch word *potaschen* (*pot ashes*, 1477).

The old method of making potassium carbonate (K_2CO_3) was by collecting or producing wood ash (an occupation carried out by ash burners), leaching the ashes and then evaporating the resulting solution in large iron pots, leaving a white residue called *pot ash*. Approximately 10% by weight of common wood ash can be recovered as pot ash. Later, *potash* became the term widely applied to naturally occurring potassium salts and the commercial product derived from them.

The following table lists a number of potassium compounds which use the word *potash* in their traditional names:

Common name	Chemical name	Formula
Potash fertilizer	c.1942 potassium carbonate (K_2CO_3); c.1950 any one or more of potassium chloride (KCl), potassium sulfate (K_2SO_4) or potassium nitrate (KNO_3). Does *not* contain potassium oxide (K_2O), which plants do not take up. However, the amount of potassium is often reported as K_2O equivalent (that is, how much it would be if in K_2O form), to allow apples-to-apples comparison between different fertilizers using different types of potash.	
Caustic potash or potash lye	potassium hydroxide	KOH
Carbonate of potash, salts of tartar, or pearlash	potassium carbonate	K_2CO_3
Chlorate of potash	potassium chlorate	$KClO_3$
Muriate of potash	potassium chloride	KCl:NaCl (95:5 or higher)
Nitrate of potash or saltpeter	potassium nitrate	KNO_3
Sulfate of potash	potassium sulfate	K_2SO_4
Permanganate of potash	potassium permanganate	$KMnO_4$

Production

All commercial potash deposits come originally from evaporite deposits and are often buried deep below the earth's surface. Potash ores are typically rich in potassium chloride (KCl) and sodium chloride (NaCl) and are typically obtained by conventional shaft mining with the extracted ore ground into a powder. Other methods include dissolution mining and evaporation methods from brines.

In the evaporation method hot water is injected into the potash which is dissolved and then pumped to the surface where it is concentrated by solar induced evaporation. Amine reagents are then added to either the mined or evaporated solutions. The amine coats the KCl but not NaCl. Air bubbles cling to the amine + KCl and float it to the surface while the NaCl and clay sink to the bottom. The surface is skimmed for the amine + KCl which is then dried and packaged for use as a K rich fertilizer—KCl dissolves readily in water and is available quickly for plant nutrition.

Potash deposits can be found all over the world. At present deposits are being mined in Canada, Russia, China, Belarus, Israel, Germany, Chile, the United States, Jordan, Spain, the United Kingdom, Uzbekistan and Brazil.

History of Production

The first U.S. patent was issued for an improvement "in the making of Pot ash and Pearl ash by
a new Apparatus and Process"; it was signed by then President George Washington.

Potash (especially potassium carbonate) has been used in bleaching textiles, making glass, and making soap, since about AD 500. Potash was principally obtained by leaching the ashes of land and sea plants. Beginning in the 14th century potash was mined in Ethiopia. One of the world's largest deposits, 140 to 150 million tons, is located in the Tigray's Dallol area. Potash was one of the most important industrial chemicals in Canada. It was refined from the ashes of broadleaved trees and produced primarily in the forested areas of Europe, Russia, and North America. The first U.S. patent of any kind was issued in 1790 to Samuel Hopkins for an improvement "in the making of Pot ash and Pearl ash by a new Apparatus and Process". *Pearl ash* was a purer quality made by the ignition of cream of tartar. Potash pits were once used in England to produce potash that was used in making soap for the preparation of wool for yarn production.

As early as 1767, potash from wood ashes was exported from Canada, and exports of potash and pearl ash (potash and lime) reached 43,958 barrels in 1865. There were 519 asheries in operation in 1871. The industry declined in the late 19th century when large-scale production of potash from mineral salts was established in Germany. In 1943, potash was discovered in Saskatchewan, Canada, in the process of drilling for oil. Active exploration began in 1951. In 1958, the Potash Company of America became the first potash producer in Canada with the commissioning of an underground potash mine at Patience Lake; however, due to water seepage in its shaft, production stopped late in 1959 but following extensive grouting and repairs, resumed in 1965. The underground mine was flooded in 1987 and was reactivated for commercial production as a solution mine in 1989.

In the late 18th and early 19th centuries, potash production provided settlers in North America a way to obtain badly needed cash and credit as they cleared wooded land for crops. To make full use of their land, settlers needed to dispose of excess wood. The easiest way to accomplish this was to burn any wood not needed for fuel or construction. Ashes from hardwood trees could then be used to make lye, which could either be used to make soap or boiled down to produce valuable potash. Hardwood could generate ashes at the rate of 60 to 100 bushels per acre (500 to 900 m³/km²). In 1790, ashes could be sold for $3.25 to $6.25 per acre ($800 to $1,500/km²) in rural New York State – nearly the same rate as hiring a laborer to clear the same area. Potash making became a major industry in British North America. Great Britain was always the most important market. The American potash industry followed the woodsman's ax across the country. After about 1820, New York replaced New England as the most important source; by 1840 the center was in Ohio. Potash production was always a by-product industry, following from the need to clear land for agriculture.

Potash evaporation ponds near Moab, Utah

Most of the world reserves of potassium (K) were deposited as sea water in ancient inland oceans. After the water evaporated, the potassium salts crystallized into beds of potash ore. These are the locations where potash is being mined today. The deposits are a naturally occurring mixture of potassium chloride (KCl) and sodium chloride (NaCl), more commonly known as table salt. Over time, as the surface of the earth changed, these deposits were covered by thousands of feet of earth.

Most potash mines today are deep shaft mines as much as 4,400 feet (1,400 m) underground. Others are mined as strip mines, having been laid down in horizontal layers as sedimentary rock. In above-ground processing plants, the KCl is separated from the mixture to produce a high-analysis potassium fertilizer. Other potassium salts can be separated by various procedures, resulting in potassium sulfate and potassium-magnesium sulfate.

Today some of the world's largest known potash deposits are spread all over the world from Saskatchewan, Canada, to Brazil, Belarus, Germany, and the Permian Basin. The Permian basin deposit includes the major mines outside of Carlsbad, New Mexico, to the world's purest potash deposit in Lea County, New Mexico (not far from the Carlsbad deposits), which is believed to be roughly 80% pure. (Osceola County, Michigan has deposits 90+% pure, however, the only mine there was recently converted to salt production.) Canada is the largest producer, followed by Russia and Belarus. The most significant reserve of Canada's potash is located in the province of Saskatchewan and controlled by the Potash Corporation of Saskatchewan.

In the beginning of the 20th century, potash deposits were found in the Dallol Depression in Musely and Crescent localities near the Ethiopean-Eritrean border. The estimated reserves are 173 and 12 million tonnes for the Musely and Crescent, respectively. The latter is particularly suitable for surface mining; it was explored in the 1960s but the works stopped due to the flood in 1967. Attempts to continue mining in the 1990s were halted by the Eritrean–Ethiopian War and have not resumed as of 2009.

Recently, recovery of potassium fertilizer salts from sea water has been studied in India. During extraction of salt from seawater by evaporation, potassium salts get concentrated in bittern, an effluent from the salt industry.

In 2013, almost 70% of potash production was controlled by two cartels, Canpotex and the Belarusian Potash Company. The latter was a joint venture between Belaruskali and Uralkali, but on July 30, 2013 Uralkali announced that it had ended the venture.

Consumption

Production and resources of potash (2011, in million tonnes)		
Country	Production	Reserves
Canada	11.2	4400
Russia	7.4	3300
Belarus	5.5	750
Germany	3.3	150
China	3.2	210
Israel	2.0	40
Jordan	1.4	40
United States	1.1	130
Chile	0.8	70
Uzbekistan	0.62	100
United Kingdom	0.43	22
Spain	0.42	20
Brazil	0.4	300
Other countries		50
World total	37.62	9600

Fertilizers

Potassium is the third major plant and crop nutrient after nitrogen and phosphorus. It has been used since antiquity as a soil fertilizer (about 90% of current use). Elemental potassium does not occur in nature because it reacts violently with water. As part of various compounds, potassium makes up about 2.6% of the Earth's crust by mass and is the seventh most abundant element, similar in abundance to sodium at approximately 1.8% of the crust. Potash is important for agriculture because it improves water retention, yield, nutrient value, taste, color, texture and disease resistance of food crops. It has wide application to fruit and vegetables, rice, wheat and other grains, sugar, corn, soybeans, palm oil and cotton, all of which benefit from the nutrient's quality-enhancing properties.

Demand for food and animal feed has been on the rise since 2000. The United States Department of Agriculture's Economic Research Service (ERS) attributes the trend to average annual population increases of 75 million people around the world. Geographically, economic growth in Asia and Latin America greatly contributed to the increased use of potash-based fertilizer. Rising incomes in developing countries also was a factor in the growing potash and fertilizer use. With more money in the household budget, consumers added more meat and dairy products to their diets. This shift in eating patterns required more acres to be planted, more fertilizer to be applied and more animals to be fed—all requiring more potash.

After years of trending upward, fertilizer use slowed in 2008. The worldwide economic downturn is the primary reason for the declining fertilizer use, dropping prices, and mounting inventories.

The world's largest consumers of potash are China, the United States, Brazil, and India. Brazil imports 90% of the potash it needs.

Potash imports and exports are often reported in K_2O equivalent, although fertilizer never contains potassium oxide, per se, because potassium oxide is caustic and hygroscopic.

Pricing

At the beginning of 2008, potash prices started a meteoric climb from less than US$200 a tonne to a high of US$875 in February 2009. These subsequently dropped dramatically to an April 2010 low of US$310 level, before recovering in 2011–12, and relapsing again in 2013. For reference, prices in November 2011 were about US$470 per tonne, but as of May 2013 were stable at US$393. After the surprise breakup of the world's largest potash cartel at the end of July 2013, potash prices are poised to drop some 20 percent. At the end of Dec 2015, potash traded for US$295 a tonne. In April 2016 its price was US$269.

Other uses

In addition to its use as a fertilizer, potassium chloride is important in many industrialized economies, where it is used in aluminium recycling, by the chloralkali industry to produce potassium hydroxide, in metal electroplating, oil-well drilling fluid, snow and ice melting, steel heat-treating, in medicine as a treatment for hypokalemia, and water softening. Potassium hydroxide is used for industrial water treatment and is the precursor of potassium carbonate, several forms of potassium phosphate, many other potassic chemicals, and soap manufacturing. Potassium carbonate is used to produce animal feed supplements, cement, fire extinguishers, food products, photographic chemicals, and textiles. It is also used in brewing beer, pharmaceutical preparations, and as a catalyst for synthetic rubber manufacturing. These non-fertilizer uses have accounted for about 15% of annual potash consumption in the United States.

Potash (potassium carbonate) along with hartshorn was also used as a baking aid similar to baking soda in old German baked goods such as lebkuchen, or gingerbread.

Granular Potash Materials

Compaction/ granulation is now an important stage in the overall potassium chloride production/beneficiation process. Compaction implies the agglomeration of particles under force to produce a densified and coherent sheet like material referred to as flake. In this case , granulation refers to the dry milling of flake. In the case, granulation refers to the dry milling of flake into a prescribed size distribution employing a screening step.

The general flowsheet of potash compaction/granulation is shown in figure below. The system is composed of four groups of equipment for the following process steps.

- Storage and feeding.
- Compaction.
- Size reduction and classification
- Finishing.

Storage and Feeding: The section usually includes:

- Material storage hopper allowing the undersize material from the granulation section to be recycled.
- Feeding hopper system, possibly with magnet and rough screening system

Compaction – The section usually includes:

- Feeder
- Compacting rolls
- Flake breaker.

There are two types of feeders: gravity and force type feeders.

A gravity system is composed of a slightly diverging chute located above the rolls and an adjustable vertical feed control tongue to keep the level constant.

The force feeders are equipped with single or multiple screws. These screws may be positioned vertically, at an angle, or horizontally. Tapered screws deaerate and predensify the feed to the compactor. Force feeders facilitate automation of the process by regulation of the speed of the screw. Both gravity and force feeders are deaerated through ports connected to the dust system.

The amount of the air removed depends on the size of material and compression ratio. In case of compacting potash from specific gravity of $1g/cm^3$ to $2g/cm^3$, the amount of air is about $0.5Nm^3$ / t of product.

The compactors are two counter rotating rolls. One roller is located in a set of fixed bearings; the other is a floating-bearing unit. The feed passed between the two rollers is progressively transformed from a loose to a dense state.

For compaction the general rule is that the larger particles should be accompanied by fines to fill in the larger pores or open space. The van der Waals attraction forces are often not sufficient to produce flake with adequate strength. Sometimes binders are added, or creation of crystalline bridges is used.

The temperature of potash feed determines its plasticity. In comparison with colder feed the warmer feed may require less pressure to obtain product with a given strength and density, or at the same energy input, produce material with greater strength and density.

The product from the compaction process is the flake. After compaction the flake is occurred in the conveying system. The flake is broken into smaller pieces immediately below the roller. Pin-type breaker and coarse tooth roll crushers are used to break the flakes.

Granulation- the granulation section is essentially composed of crushing and screening equipment. Potash granulation mainly consists of primary and secondary crushers in closed circuit with a multi-deck screening system. Two stage milling decreases the recycle ratio. The primary crusher grinds large materials from the compactor, and the secondary crusher is fed with oversize from the screen. Hammer mills, cage mills. Or chain mills are used as the primary crushers. Roll crushers with toothed rolls are often used as secondary crushers.

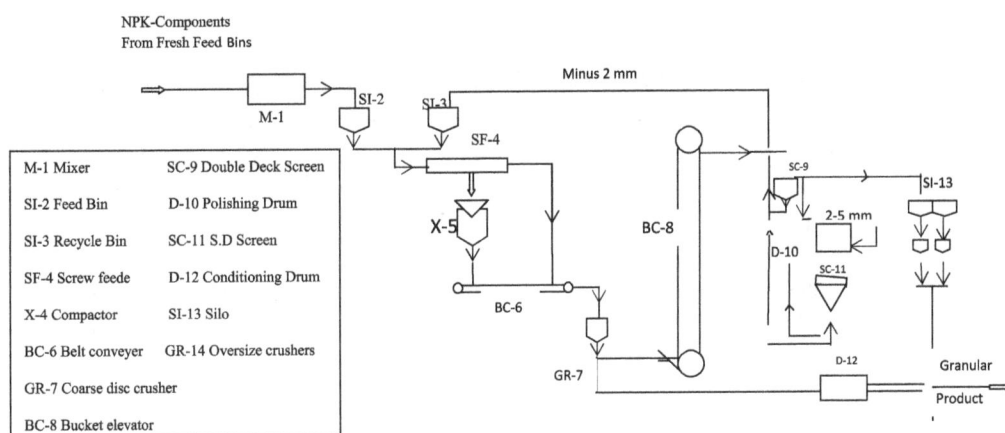

General Flowsheet for Potash Compaction/Granulation Plant

Finishing Section

- Polishing unit
- Coating unit
- Final product storage.

The particles after granulation have sharp edges and are of irregular shape. To diminish the crushing of the material during further handling and dust creation the potash particles are quenched, dried, and passed through the polishing screen. Quenching consists of wetting particles with water or brine by direct spraying on the conveyer or mixing in screw conveyer. During the quenching stage the sharper corners of the particles break off and a shell of dissolved salt envelops the particles. After wetting the product is dried in a rotary cirum or fluid bed dryer at a temperature as high as 200°C. Dry product is passed over a polishing screen to remove the fines.

In the final steps the product is cooled and coated in a rotary drum with anticaking agents (amines and oils). Amines depress the caking tendency of potash during transportations and oils prevent the formation of dust clouds at the heading points.

Dust Collection System- Major dust sources are the hopper in which the payloader dumps the raw material, the crushers, the transfer points, the mixer, and the screens. The amount of air needed in the dust collection system depends on the number and types of collection points. The system consists of a fan, cyclones, and filter bags connected by air ducts.

Potassium Sulfate

Potassium sulfate is the second largest tonnage potassium compound and it is also used primarily as a fertilizer. The sulfate or other nonchloride forms of potassium are preferred for certain crops that do not tolerate the chloride ion well, eg., tobacco and some ffruits and vegetable. Nonchloride potash sources are also needed in areas where chloride accumulate or in areas of very intensive agriculture, potassium sulfate may be preferred because of its sulfur content where soils are deficient in both potassium and sulfur.

Potassium Sulfate Production

Mannheim Process – Historically potassium sulfate has been made primarily from KCL and sulfuric acid (and a small amount from KCL and SO_2) when the byproduct HCL was the dominant product. However, over the year the HCL market has had more competition and "natural" K_2SO_4 with lower capital and operating coasts has begun to dominate its production in some countries with natural complex salts.

The Mannheim process was originally developed from sodium sulfate production by reacting NaCl with sulfuric acid. Replacing NaCl with KCl produces potassium sulfate. The reaction is two-stage:

a. Exothermic reaction

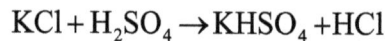

$$KCl + H_2SO_4 \rightarrow KHSO_4 + HCl$$

b. Endothermic reaction

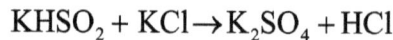

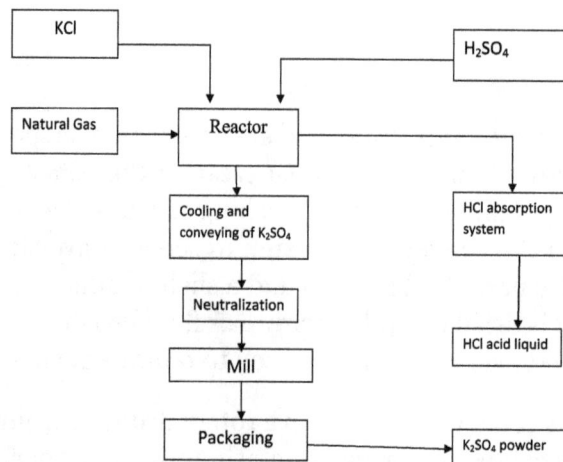

$$KHSO_2 + KCl \rightarrow K_2SO_4 + HCl$$

Schematic diagram of Potassium Sulphate production

The potassium chloride reacts during slow mixing in the heated Mannheim furnace with sulfuric acid, producing gaseous HCl and K_2SO_4. The furnace is heated by natural gas or fuel oil. The product K_2SO_4 is cooled in a cooling drum. Lump material from the cooler is crushed and finished or can be compacted and granulated as with KCl.

The HCl gas is cooled in a graphite in a graphite heat exchanger and absorbed in water in two stages to produce 30% hydrochloride acid as a byproduct. The process gives an excellent quality that contains over 50% K_2O and less than 1% chlorine. Emissions are well controlled.

Recovery of Potassium Sulfate From Natural Complex Salts – The chief natural complex salts that are the source of the potassium sulfate are;

- Kainite ($KCl . MgSO_4 . 3H_2O$)
- Langbeinite ($K_2SO_4 . 3MgSO_4$)
- Carpathian polmineral ores

The natural process involves the initial conversion with recycled K_2SO_4 end liquor "mined" kainite or langbeinite to form an intermediate product schoenite. All processes are based on intercrystalline reactions of ion exchange.

The Process comprises four basic units:

- Preparation of the ore and flotation;
- Production of schoenite and its recovery;
- Leaching of the schoenite to potassium sulfate;
- Liquor treatment.

The kainite is repulped with recycled brine, screened, and directed to ball mills and hydroclassifiers. Overflows go to a thicker and main filter and underflows to flotation and filtration. Float material, after filtration, is combined with the solid fraction from the main filter and directed to the schoenite reactors and separating cyclones. After a two step hydroseparation, centrifugation, and filtration, schoenite is directed to the to the leaching reactors. After decomposition of the schoenite, product is directed to final centrifuges and a dryer; the overflows are cooled and crystallized. After additional thickening, the product is centrifuged and dried. The product specification ensures that the K_2O content is not lower than 50% and the chlorine content is less than 1%.

In the recovery unit, italkali forms syngenite to moderately increase the recovery of schoenite from the plant end liquor. The brine from the schoenite filter is reacted with main flow to be leaching reactors.

Other processes involve adding sylvite to kainite, langbeinite kieserite, etc. The schoenite intermediate can be formed kieserite, etc. the schoenite intermediate can be formed by reacting KCl with mined kieserite or the epsomite. Where solar or plant evaporation can be done economically, the yields can be further improved by evaporating the schoenite or glaserite end liquor and recycling the salts.

A complex process can production of potassium sulfate from ores can been implemented on an industrial scale. The naturals ores are composed of calcium sulfate anhydride, epsomite, halite, kainite, kieserite, langbeinite, polyhalite, sylvite and clay. The treatment of such an ore requires permanent analytical services and the development of a large number and for the kind of salts intended to be produced, additional sylvite must be added in varying amounts to maintain the right proportions for crystal formation. The basic process concept is to produce schoenite from all available salts. The remaining products or processes are subjugated to his basic process.

The Carpathian ore contains about 9% potassium and 15% clay. The ore is leached with hot synthetic kainite solution in a dissolution chamber. The langbeinite, polyhalite and halite remain undissolved chamber containing salts and clay is directed to a Dorr-Oliver settler where clay is settled and directed to a washer and discarded. The solution is crystallized at the proper cation and anion proportions to produce crystalline schoenite. To avoid crystalline of potassium chloride and sodium chloride, the saturated solution of potassium and magnesium sulfates is added to the Dorr-Oliver settler. The slurry of schoenite is filtered and crystals are leached with water to produces K_2SO_4 crystals, which are centrifuged and recycled and a liquor of potassiumand magnesium sulfates. Also liquid phase from filter is recycled and added to the schoenite liquor from vacuum crystallization. Part of the schoenite liquor is evaporated to produce crystalline sodium sulfate and discard the magnesium chloride liquid end products. The slurry from the evaporation unit is recycled as "synthetic kainite". This process permits the use of the Carpathian ores to produce several commercially valuable products such as potassium sulfate, potassium-magnesium sulfate, potassium chloride, sodium sulfate and magnesium chloride liquors. Neither the economic evaluation of the process nor any of the consumption figures has been published.

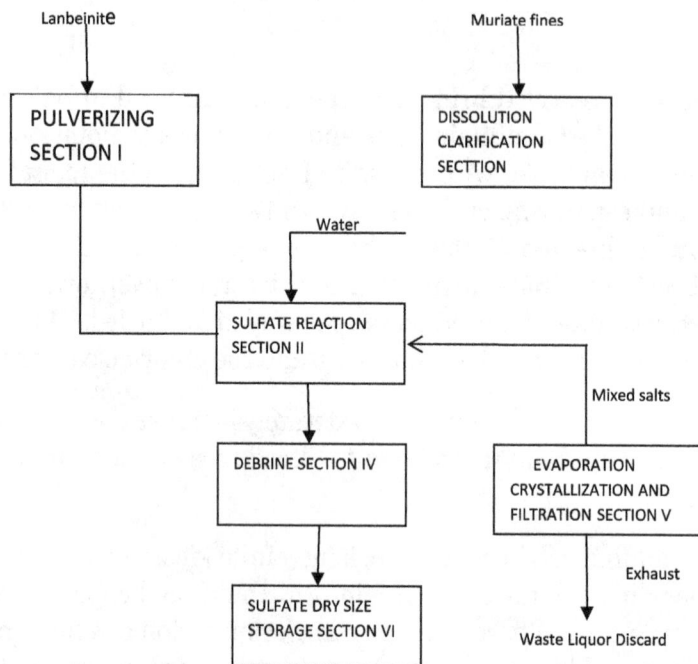

Potassium Sulfate Process from Langbeinite

The production of potassium sulfate from langbeinite is possible with a large amount of muriate of potash is possible with a large amount of nitrate of potash by mixing langbeinite and sylvite.

The langbeinite ore is separated from sylvite and halite by selective washing, froth flotation, or heavy media separation. The commercial langbeinite used in the process must pulverized in ball mills, and fine powder is mixed with a solution of the muriate of potash. The muriate of potash is dissolved and clarified in a separate unit. The reaction in the presence of water yield potassium sulfate in a crystalline form and brine. Crystals are centrifuged or filtered, dried in a rotary dryer, sized and finished. The finished methods either produce coarse material or granulated product. The mixed salts are added to the sulfate reactor the liquor is discard as a waste.

Potassium Nitrate

Potassium nitrate is a chemical compound with the chemical formula KNO_3. It is an ionic salt of potassium ions K^+ and nitrate ions NO_3^-, and is therefore an alkali metal nitrate.

It occurs as a mineral niter and is a natural solid source of nitrogen. Potassium nitrate is one of several nitrogen-containing compounds collectively referred to as saltpeter or saltpetre.

Major uses of potassium nitrate are in fertilizers, tree stump removal, rocket propellants and fireworks. It is one of the major constituents of gunpowder (black powder) and has been used since the Middle Ages as a food preservative.

Etymology

Potassium nitrate, because of its early and global use and production, has many names.

The Greeks used the term *nitron*, which was Latinised to *nitrum* or *nitrium,* while earlier Hebrews and Egyptians used words with the consonants n-t-r, which leads some to speculate that the Latin term is closer to the original than the Greek term. Middle English styled it *nitre*. Old French had *niter*. By the 15th century, Europeans referred to it as *saltpeter* and later as *nitrate of potash,* as the chemistry of the compound was more fully understood.

Properties

Potassium nitrate has an orthorhombic crystal structure at room temperature, which transforms to a trigonal system at 129 °C (264 °F).

Potassium nitrate is moderately soluble in water, but its solubility increases with temperature. The aqueous solution is almost neutral, exhibiting pH 6.2 at 14 °C (57 °F) for a 10% solution of commercial powder. It is not very hygroscopic, absorbing about 0.03% water in 80% relative humidity over 50 days. It is insoluble in alcohol and is not poisonous; it can react explosively with reducing agents, but it is not explosive on its own.

Thermal Decomposition

Between 550–790 °C (1,022–1,454 °F), potassium nitrate reaches a temperature dependent equilibrium with potassium nitrite:

$$2\ KNO_3 \rightleftharpoons 2\ KNO_2 + O_2$$

History of Production

From Mineral Sources

The earliest known complete purification process for potassium nitrate was outlined in 1270 by the chemist and engineer Hasan al-Rammah of Syria in his book *al-Furusiyya wa al-Manasib al-Harbiyya* (*The Book of Military Horsemanship and Ingenious War Devices*). In this book, al-Rammah describes first the purification of *barud* (crude saltpeter mineral) by boiling it with minimal water and using only the hot solution, then the use of potassium carbonate (in the form of wood ashes) to remove calcium and magnesium by precipitation of their carbonates from this solution, leaving a solution of purified potassium nitrate, which could then be dried. This was used for the manufacture of gunpowder and explosive devices. The terminology used by al-Rammah indicated a Chinese origin for the gunpowder weapons about which he wrote.

At least as far back as 1845, Chilean saltpeter deposits were exploited in Chile and California, USA.

From Caves

A major natural source of potassium nitrate was the deposits crystallizing from cave walls and the accumulations of bat guano in caves. Extraction is accomplished by immersing the guano in water for a day, filtering, and harvesting the crystals in the filtered water. Traditionally, guano was the source used in Laos for the manufacture of gunpowder for *Bang Fai* rockets.

LeConte

Perhaps the most exhaustive discussion of the production of this material is the 1862 LeConte text. He was writing with the express purpose of increasing production in the Confederate States to support their needs during the American Civil War. Since he was calling for the assistance of rural farming communities, the descriptions and instructions are both simple and explicit. He details the "French Method", along with several variations, as well as a "Swiss method". N.B. Many references have been made to a method using only straw and urine, but there is no such method in this work.

French Method

Niter-beds are prepared by mixing manure with either mortar or wood ashes, common earth and organic materials such as straw to give porosity to a compost pile typically 4 feet (1.2 m) high, 6 feet (1.8 m) wide, and 15 feet (4.6 m) long. The heap was usually under a cover from the rain, kept moist with urine, turned often to accelerate the decomposition, then finally leached with water after approximately one year, to remove the soluble calcium nitrate which was then converted to potassium nitrate by filtering through the potash.

Swiss Method

LeConte describes a process using only urine and not dung, referring to it as the *Swiss method*. Urine is collected directly, in a sandpit under a stable. The sand itself is dug out and leached for nitrates which were then converted to potassium nitrate via potash, as above.

From Nitric Acid

From 1903 until the World War I era, potassium nitrate for black powder and fertilizer was produced on an industrial scale from nitric acid produced via the Birkeland–Eyde process, which used an electric arc to oxidize nitrogen from the air. During World War I the newly industrialized Haber process (1913) was combined with the Ostwald process after 1915, allowing Germany to produce nitric acid for the war after being cut off from its supplies of mineral sodium nitrates from Chile.

Production

Potassium nitrate can be made by combining ammonium nitrate and potassium hydroxide.

$$NH_4NO_3 \text{ (aq)} + KOH \text{ (aq)} \rightarrow NH_3 \text{ (g)} + KNO_3 \text{ (aq)} + H_2O \text{ (l)}$$

An alternative way of producing potassium nitrate without a by-product of ammonia is to combine ammonium nitrate and potassium chloride, easily obtained as a sodium-free salt substitute.

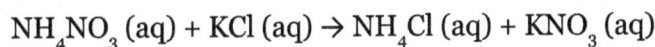

$$NH_4NO_3 \text{ (aq)} + KCl \text{ (aq)} \rightarrow NH_4Cl \text{ (aq)} + KNO_3 \text{ (aq)}$$

Potassium nitrate can also be produced by neutralizing nitric acid with potassium hydroxide. This reaction is highly exothermic.

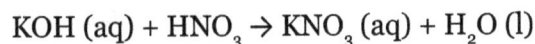

$$KOH \text{ (aq)} + HNO_3 \rightarrow KNO_3 \text{ (aq)} + H_2O \text{ (l)}$$

On industrial scale it is prepared by the double displacement reaction between sodium nitrate and potassium chloride.

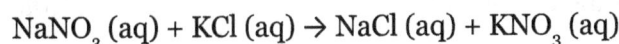

$$NaNO_3 \text{ (aq)} + KCl \text{ (aq)} \rightarrow NaCl \text{ (aq)} + KNO_3 \text{ (aq)}$$

Uses

Potassium nitrate has a wide variety of uses, largely as a source of nitrate.

Nitric Acid Production

Historically, nitric acid was produced by combining sulfuric acid with nitrates such as saltpeter. In modern times this is reversed: nitrates are produced from nitric acid produced via the Ostwald process.

Oxidizer

The most famous use of potassium nitrate is probably as the oxidizer in blackpowder. From the most ancient times through the late 1880s, blackpowder provided the explosive power for all the world's firearms. After that time, small arms and large artillery increasingly began to depend on cordite, a smokeless powder. Blackpowder remains in use today in black powder rocket motors, but also in combination with other fuels like sugars in "rocket candy". It is also used in fireworks such as smoke bombs. It is also added to cigarettes to maintain an even burn of the tobacco and is used to ensure complete combustion of paper cartridges for cap and ball revolvers. It can also be heated to several hundred degrees to be used for niter bluing, which is less durable than other forms of protective oxidation, but allows for specific and often beautiful coloration of steel parts, such as screws, pins, and other small parts of firearms.

Food Preservation

In the process of food preservation, potassium nitrate has been a common ingredient of salted meat since the Middle Ages, but its use has been mostly discontinued because of inconsistent results compared to more modern nitrate and nitrite compounds. Even so, saltpeter is still used in some food applications, such as *charcuterie* and the brine used to make corned beef. When used as a food additive in the European Union, the compound is referred to as E252; it is also approved for use as a food additive in the USA and Australia and New Zealand (where it is listed under its INS number 252). Although nitrate salts have been suspected of producing the carcinogen nitrosamine, both sodium and potassium nitrates and nitrites have been added to meats in the US since 1925, and nitrates and nitrites have not been removed from preserved meat products because nitrite and nitrate inhibits the germination of C. botulinum endospores, and thus prevents botulism from bacterial toxin that may otherwise be produced in certain preserved meat products.

Food Preparation

In West African cuisine, potassium nitrate (saltpetre) is widely used as a thickening agent in soups and stews such as okra soup and isi ewu. It is also used to soften food and reduce cooking time when boiling beans and tough meat. Saltpetre is also an essential ingredient in making special porridges, such as *kunun kanwa* literally translated from the Hausa language as 'saltpetre porridge'. In the Shetland Islands (UK) it is used in the curing of mutton to make "reestit" mutton, a local delicacy.

Fertilizer

Potassium nitrate is used in fertilizers as a source of nitrogen and potassium – two of the macro-nutrients for plants. When used by itself, it has an NPK rating of 13-0-44.

Pharmacology

- Used in some toothpastes for sensitive teeth. Recently, the use of potassium nitrate in toothpastes for treating sensitive teeth has increased and it may be an effective treatment.

- Used historically to treat asthma. Used in some toothpastes to relieve asthma symptoms.

- Used in Thailand as main ingredient in kidney tablets to relieve the symptoms of cystitis, pyelitis and urethritis.

- Combats high blood pressure and was once used as a hypotensive.

Other uses

- Electrolyte in a salt bridge

- Active ingredient of condensed aerosol fire suppression systems. When burned with the free radicals of a fire's flame, it produces potassium carbonate.

- Works as an aluminium cleaner.

- Component (usually about 98%) of some tree stump removal products. It accelerates the natural decomposition of the stump by supplying nitrogen for the fungi attacking the wood of the stump.

- In heat treatment of metals as a medium temperature molten salt bath, usually in combination with sodium nitrite. A similar bath is used to produce a durable blue/black finish typically seen on firearms. Its oxidizing quality, water solubility, and low cost make it an ideal short-term rust inhibitor.

- To induce flowering of mango trees in the Philippines.

- Thermal storage medium in power generation systems. Sodium and potassium nitrate salts are stored in a molten state with the solar energy collected by the heliostats at the Gemasolar Thermosolar Plant. Ternary salts, with the addition of calcium nitrate or lithium nitrate, have been found to improve the heat storage capacity in the molten salts.

In Folklore and Popular Culture

Potassium nitrate was once thought to induce impotence, and is still falsely rumored to be in institutional food (such as military fare) as an anaphrodisiac; however, there is no scientific evidence for such properties.

Compound Fertilizers

Potash is often applied in mixtures with other nutrients to provide the specific fertilizers needed by crops or soil and to allow placement in one application. To meet this need most fertilizer dealers have modest blending equipment, but to prevent segregation the fertilizers must be in the granular or coarse form.

About 4%-5% of potash production is used in industrial applications. The industrial potash (chemical grade) has a different purity from fertilizer grade potash. The composition of chemical grade potash is given. Product is shippedin bags or in bulk form in modified hopper cars called spargers (in slurry form) or solution cars (in liquid form).

Chemical-grade potash has the following consumption pattern:

Detergents and soaps	35%-30%
Glass and ceramics	25%-28%
Textiles and dyes	20%-22%

Chemicals and drugs	13%-15%
Other	7%-5%

Most of the chemical grade potash is used for production of potassium hydroxide. Potassium hydroxide (KOH) and its derivative potassium carbonate are the next largest industrial potassium compounds. The KOH is made by the electrolysis of KCL in installations similar to caustic soda/chlorine production. The main use of caustic potash is in the manufacture of liquid soaps; textile operations; production of grease, catalysts, alkaline batteries electro polishing and rubber production.

Several other potassium compounds have a limited use in agriculture but much wider use in industrial or commercial applications. Potassium phosphate, for instance, is used in some high analysis, low salt content fertilizer mixtures. It is not yet a large tonnage fertilizer but several companies over the years have announced plans to produce it for that purpose. Potash is reacted with sulfuric Acid, and the HCl is removed at the $KHSO_4$ stage. This salt is then reacted with more H_2SO_4 and phosphate rock $[Ca_3(PO_4)_2]$ usually to produce the more easily crystallized monopotassium phosphate (KH_2PO_4). However, the yields are low because of K losses with many of the impurities in the rock and the process is corrosive and complex. By far the widest use of potassium phosphate has been as an additive to heavy duty detergent, predominantly as low molecular weight polymers created by fusing various KOH and H_3PO_4 mixtures.

Potassium carbonate is used primarily in the glass manufacture eg., of television and similar display tubes. Potassium carbonate solution (0-0-30 grade) is marketed in the United States as a specialty liquid fertilizer. Potassium carbonate is produced by carbonating KOH with CO_2. Some potassium carbonate is further carbonated to produce $KHCO_3$ (potassium bicarbonate), which is used largely in the food and pharmaceutical industries.

References

- B. J. Kosanke; B. Sturman; K. Kosanke; I. von Maltitz; T. Shimizu; M. A. Wilson; N. Kubota; C. Jennings-White; D. Chapman (2004). "2". Pyrotechnic Chemistry. Journal of Pyrotechnics. pp. 5–6

- Arnold F. Holleman, Egon Wiberg and Nils Wiberg (1985). "Potassium". Lehrbuch der Anorganischen Chemie (in German) (91–100 ed.). Walter de Gruyter. ISBN 3-11-007511-3

- Joseph LeConte (1862). Instructions for the Manufacture of Saltpeter. Columbia, S.C.: South Carolina Military Department. p. 14. Retrieved 2007-10-19

- Enomoto, K; et al. (2003). "The Effect of Potassium Nitrate and Silica Dentifrice in the Surface of Dentin". Japanese Journal of Conservative Dentistry. 46 (2): 240–247

- J. W. Turrentine (1934). "Composition of Potash Fertilizer Salts for Sale on the American Market". Industrial & Engineering Chemistry. American Chemical Society. 26 (11): 1224–1225. doi:10.1021/ie50299a022

- Oliver Frederick Gillilan Hogg (1993). Clubs to cannon: warfare and weapons before the introduction of gunpowder (reprint ed.). Barnes & Noble Books. p. 216. ISBN 1-56619-364-8. Retrieved 2011-11-28

- Juan Ignacio Burgaleta; Santiago Arias; Diego Ramirez. "Gemasolar, The First Tower Thermosolar Commercial Plant With Molten Salt Storage System" (PDF) (Press Release). Torresol Energy. Archived (PDF) from the original on 9 March 2012. Retrieved 7 March 2012

- R. Orchardson & D. G. Gillam (2006). "Managing dentin hypersensitivity" (PDF). Journal of the American Dental Association (1939). 137 (7): 990–8; quiz 1028–9. PMID 16803826. doi:10.14219/jada.archive.2006.0321

- Partington, J. R. (1960). A History of Greek Fire and Gunpowder (illustrated, reprint ed.). JHU Press. p. 335. ISBN 0801859549. Retrieved 2014-11-21

- van der Sijs i.a., Nicoline (2010). "POTAS (SCHEIKUNDIG ELEMENT)". Etymologiebank (in Dutch). Retrieved 14 August 2016

- Hall, William L; Robarge, Wayne P; Meeting, American Chemical Society (2004). Environmental Impact of Fertilizer on Soil and Water. p. 40. ISBN 9780841238114

- Major George Rains (1861). Notes on Making Saltpetre from the Earth of the Caves. New Orleans, LA: Daily Delta Job Office. p. 14. Retrieved September 13, 2012

- Richard E. Jones & Kristin H. López (2006). Human Reproductive Biology, Third Edition. Elsevier/Academic Press. p. 225. ISBN 0-12-088465-8

- Eli S. Freeman (1957). "The Kinetics of the Thermal Decomposition of Potassium Nitrate and of the Reaction between Potassium Nitrite and Oxygen". J. Am. Chem. Soc. 79 (4): 838–842. doi:10.1021/ja01561a015

Impact of Fertilizers on Environment

The impact of fertilizers on the environment depends on the agronomic practices of the area. Some of the environmental issues related to the issue of the environment are deforestation, irrigation problems, soil degradation and waste. Eutrophication, soil contamination, soil salinity and human impact on the nitrogen cycle are some of the topic discussed in the following chapter. Fertilizers are best understood in confluence with the major topics listed in the following chapter.

Groundwater Pollution

Groundwater pollution example in Lusaka, Zambia where the pit latrine in the background is polluting the shallow well in the foreground with pathogens and nitrate.

Groundwater pollution (also called groundwater contamination) occurs when pollutants are released to the ground and make their way down into groundwater. It can also occur naturally due to the presence of a minor and unwanted constituent, contaminant or impurity in the groundwater, in which case it is more likely referred to as contamination rather than pollution.

The pollutant creates a contaminant plume within an aquifer. Movement of water and dispersion within the aquifer spreads the pollutant over a wider area. Its advancing boundary, often called a plume edge, can intersect with groundwater wells or daylight into surface water such as seeps and spring, making the water supplies unsafe for humans and wildlife. The movement of the plume, called a plume front, may be analyzed through a hydrological transport model or groundwater model. Analysis of groundwater pollution may focus on soil characteristics and site geology, hydrogeology, hydrology, and the nature of the contaminants.

Pollution can occur from on-site sanitation systems, landfills, effluent from wastewater treatment plants, leaking sewers, petrol filling stations or from over application of fertilizers in agriculture. Pollution (or contamination) can also occur from naturally occurring contaminants, such as arsenic or fluoride. Using polluted groundwater causes hazards to public health through poisoning or the spread of disease.

Different mechanisms have influence on the transport of pollutants, e.g. diffusion, adsorption, precipitation, decay, in the groundwater. The interaction of groundwater contamination with surface waters is analyzed by use of hydrology transport models.

Fertilizers and Pesticides

Nitrate can also enter the groundwater via excessive use of fertilizers, including manure spreading. This is because only a fraction of the nitrogen-based fertilizers is converted to produce and other plant matter. The remainder accumulates in the soil or lost as run-off. High application rates of nitrogen-containing fertilizers combined with the high water-solubility of nitrate leads to increased runoff into surface water as well as leaching into groundwater, thereby causing groundwater pollution. The excessive use of nitrogen-containing fertilizers (be they synthetic or natural) is particularly damaging, as much of the nitrogen that is not taken up by plants is transformed into nitrate which is easily leached.

Poor management practices in manure spreading can introduce both pathogens
|and nutrients (nitrate) in the groundwater system.

The nutrients, especially nitrates, in fertilizers can cause problems for natural habitats and for human health if they are washed off soil into watercourses or leached through soil into groundwater. The heavy use of nitrogenous fertilizers in cropping systems is the largest contributor to anthropogenic nitrogen in groundwater worldwide.

Feedlots/animal corrals can also lead to the potential leaching of nitrogen and metals to groundwater. Over application of animal manure may also result in groundwater pollution with pharmaceutical residues derived from veterinary drugs.

The US Environmental Protection Agency (EPA) and the European Commission are seriously dealing with the nitrate problem related to agricultural development, as a major water supply problem that requires appropriate management and governance.

Runoff of pesticides may leach into groundwater causing human health problems from contaminated water wells. Pesticide concentrations found in groundwater are typically low, and often the regulatory human health-based limits exceeded are also very low. The organophosphorus insecticide monocrotophos (MCP) appears to be one of a few hazardous, persistent, soluble and mobile (it does not bind with minerals in soils) pesticides able to reach a drinking-water source. In general, more pesticide compounds are being detected as groundwater quality monitoring programs have become more extensive; however, much less monitoring has been conducted in developing countries due to the high analysis costs.

Environmental Impact of Agriculture

Water pollution in a rural stream due to runoff from farming activity in New Zealand.

The environmental impact of agriculture is the effect that different farming practices have on the ecosystems around them, and how those effects can be traced back to those practices. The environmental impact of agriculture varies based on the wide variety of agricultural practices employed around the world. Ultimately, the environmental impact depends on the production practices of the system used by farmers. The connection between emissions into the environment and the farming system is indirect, as it also depends on other climate variables such as rainfall and temperature.

There are two types of indicators of environmental impact: "means-based", which is based on the farmer's production methods, and "effect-based", which is the impact that farming methods have on the farming system or on emissions to the environment. An example of a means-based indicator would be the quality of groundwater, that is effected by the amount of nitrogen applied to the soil. An indicator reflecting the loss of nitrate to groundwater would be effect-based. The means-based evaluation looks at farmers' practices of agriculture, and the effect-based evaluation considers the actual effects of the agricultural system. For example, means-based analysis might look at pesticides and fertilization methods that farmers are using, and effect-based analysis would consider how much CO_2 is being emitted or what the Nitrogen content of the soil is.

The environmental impact of agriculture involves a variety of factors from the soil, to water, the air, animal and soil variety, people, plants, and the food itself. Some of the environmental issues that are related to agriculture are climate change, deforestation, genetic engineering, irrigation problems, pollutants, soil degradation, and waste.

Eutrophication

The eutrophication of the Potomac River is evident from the bright green water, caused by a dense bloom of cyanobacteria.

Eutrophication, or more precisely hypertrophication, is the enrichment of a water body with nutrients, usually with an excess amount of nutrients. This process induces growth of plants and algae and due to the biomass load, may result in oxygen depletion of the water body. One example is the "bloom" or great increase of phytoplankton in a water body as a response to increased levels of nutrients. Eutrophication is almost always induced by the discharge of phosphate-containing detergents, fertilizers, or sewage, into an aquatic system.

Mechanism of Eutrophication

Eutrophication arises from the oversupply of nutrients, which leads to overgrowth of plants and algae. After such organisms die, the bacterial degradation of their biomass consumes the oxygen in the water, thereby creating the state of hypoxia.

According to Ullmann's Encyclopedia, "the primary limiting factor for eutrophication is phosphate." The availability of phosphorus generally promotes excessive plant growth and decay, favouring simple algae and plankton over other more complicated plants, and causes a severe reduction in water quality. Phosphorus is a necessary nutrient for plants to live, and is the limiting factor for plant growth in many freshwater ecosystems. Phosphate adheres tightly to soil, so it is mainly transported by erosion. Once translocated to lakes, the extraction of phosphate into water is slow, hence the difficulty of reversing the effects of eutrophication.

The sources of these excess phosphates are phosphates in detergent, industrial/domestic run-offs, and fertilizers. With the phasing out of phosphate-containing detergents in the 1970s, industrial/domestic run-off and agriculture have emerged as the dominant contributors to eutrophication.

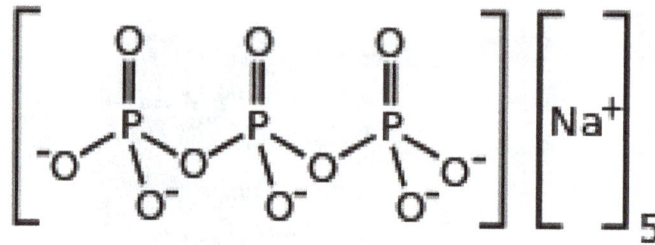

Sodium triphosphate, once a component of many detergents, was a major contributor to eutrophication.

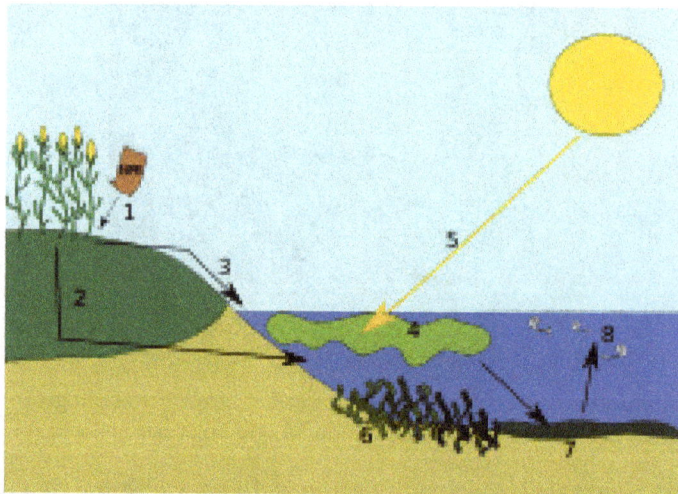

1. Excess nutrients are applied to the soil. 2. Some nutrients leach into the soil where they can remain for years. Eventually, they get drained into the water body. 3. Some nutrients run off over the ground into the body of water. 4. The excess nutrients cause an algal bloom. 5. The algal bloom blocks the light of the sun from reaching the bottom of the water body. 6. The plants beneath the algal bloom die because they cannot get sunlight to photosynthesize. 7. Eventually, the algal bloom dies and sinks to the bottom of the lake. Bacteria begins to decompose the remains, using up oxygen for respiration. 8. The decomposition causes the water to become depleted of oxygen. Larger life forms, such as fish, suffocate to death. This body of water can no longer support life.

Cultural Eutrophication

Cultural Eutrophication is the process that speeds up natural eutrophication because of human activity. Due to clearing of land and building of towns and cities, land runoff is accelerated and more nutrients such as phosphates and nitrate are supplied to lakes and rivers, and then to coastal estuaries and bays. Extra nutrients are also supplied by treatment plants, golf courses, fertilizers, farms, as well as untreated sewage in many countries.

Lakes and Rivers

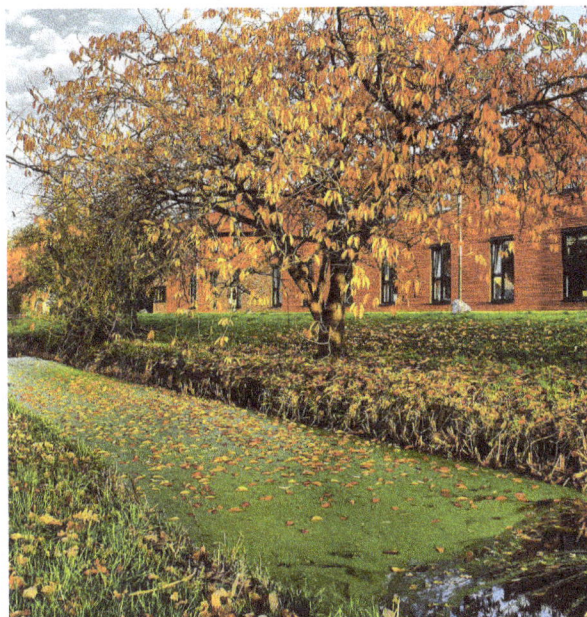

Eutrophication in a canal

When algae die, they decompose and the nutrients contained in that organic matter are converted into inorganic form by microorganisms. This decomposition process consumes oxygen, which reduces the concentration of dissolved oxygen. The depleted oxygen levels in turn may lead to fish kills and a range of other effects reducing biodiversity. Nutrients may become concentrated in an anoxic zone and may only be made available again during autumn turn-over or in conditions of turbulent flow.

Enhanced growth of aquatic vegetation or phytoplankton and algal blooms disrupts normal functioning of the ecosystem, causing a variety of problems such as a lack of oxygen needed for fish and shellfish to survive. The water becomes cloudy, typically coloured a shade of green, yellow, brown, or red. Eutrophication also decreases the value of rivers, lakes and aesthetic enjoyment. Health problems can occur where eutrophic conditions interfere with drinking water treatment.

Human activities can accelerate the rate at which nutrients enter ecosystems. Runoff from agriculture and development, pollution from septic systems and sewers, sewage sludge spreading, and other human-related activities increase the flow of both inorganic nutrients and organic substances into ecosystems. Elevated levels of atmospheric compounds of nitrogen can increase nitrogen availability. Phosphorus is often regarded as the main culprit in cases of eutrophication in lakes subjected to "point source" pollution from sewage pipes. The concentration of algae and the trophic state of lakes correspond well to phosphorus levels in water. Studies conducted in the Experimental Lakes Area in Ontario have shown a relationship between the addition of phosphorus and the rate of eutrophication. Humankind has increased the rate of phosphorus cycling on Earth by four times, mainly due to agricultural fertilizer production and application. Between 1950 and 1995, an estimated 600,000,000 tonnes of phosphorus was applied to Earth's surface, primarily on croplands. Policy changes to control point sources of phosphorus have resulted in rapid control of eutrophication.

Natural Eutrophication

Although eutrophication is commonly caused by human activities, it can also be a natural process, particularly in lakes. Eutrophy occurs in many lakes in temperate grasslands, for instance. Paleo-limnologists now recognise that climate change, geology, and other external influences are critical in regulating the natural productivity of lakes. Some lakes also demonstrate the reverse process (meiotrophication), becoming less nutrient rich with time. The main difference between natural and anthropogenic eutrophication is that the natural process is very slow, occurring on geological time scales.

Ocean Waters

Eutrophication is a common phenomenon in coastal waters. In contrast to freshwater systems, nitrogen is more commonly the key limiting nutrient of marine waters; thus, nitrogen levels have greater importance to understanding eutrophication problems in salt water. Estuaries tend to be naturally eutrophic because land-derived nutrients are concentrated where run-off enters a confined channel. Upwelling in coastal systems also promotes increased productivity by conveying deep, nutrient-rich waters to the surface, where the nutrients can be assimilated by algae.

The World Resources Institute has identified 375 hypoxic coastal zones in the world, concentrated in coastal areas in Western Europe, the Eastern and Southern coasts of the US, and East Asia, particularly Japan.

In addition to runoff from land, atmospheric fixed nitrogen can enter the open ocean. A study in 2008 found that this could account for around one third of the ocean's external (non-recycled) nitrogen supply, and up to 3% of the annual new marine biological production. It has been suggested that accumulating reactive nitrogen in the environment may prove as serious as putting carbon dioxide in the atmosphere.

Terrestrial Ecosystems

Terrestrial ecosystems are subject to similarly adverse impacts from eutrophication. Increased nitrates in soil are frequently undesirable for plants. Many terrestrial plant species are endangered as a result of soil eutrophication, such as the majority of orchid species in Europe. Meadows, forests, and bogs are characterized by low nutrient content and slowly growing species adapted to those levels, so they can be overgrown by faster growing and more competitive species. In meadows, tall grasses that can take advantage of higher nitrogen levels may change the area so that natural species may be lost. Species-rich fens can be overtaken by reed or reedgrass species. Forest undergrowth affected by run-off from a nearby fertilized field can be turned into a nettle and bramble thicket.

Chemical forms of nitrogen are most often of concern with regard to eutrophication, because plants have high nitrogen requirements so that additions of nitrogen compounds will stimulate plant growth. Nitrogen is not readily available in soil because N_2, a gaseous form of nitrogen, is very stable and unavailable directly to higher plants. Terrestrial ecosystems rely on microbial nitrogen fixation to convert N_2 into other forms such as nitrates. However, there is a limit to how much nitrogen can be utilized. Ecosystems receiving more nitrogen than the plants require are called nitrogen-saturated. Saturated terrestrial ecosystems then can contribute both inorganic and organic nitrogen to freshwater, coastal, and marine eutrophication, where nitrogen is also typically a limiting nutrient. This is also the case with increased levels of phosphorus. However, because phosphorus is generally much less soluble than nitrogen, it is leached from the soil at a much slower rate than nitrogen. Consequently, phosphorus is much more important as a limiting nutrient in aquatic systems.

Ecological Effects

Eutrophication is apparent as increased turbidity in the northern part of the Caspian Sea, imaged from orbit.

Eutrophication was recognized as a water pollution problem in European and North American lakes and reservoirs in the mid-20th century. Since then, it has become more widespread. Surveys showed that 54% of lakes in Asia are eutrophic; in Europe, 53%; in North America, 48%; in South America, 41%; and in Africa, 28%. In South Africa, a study by the CSIR using remote sensing has shown more than 60% of the dams surveyed were eutrophic. Some South African scientists believe that this figure might be higher with the main source being dysfunctional sewage works that produce more than 4 billion liters a day of untreated, or at best partially treated, sewage effluent that discharges into rivers and dams.

Many ecological effects can arise from stimulating primary production, but there are three particularly troubling ecological impacts: decreased biodiversity, changes in species composition and dominance, and toxicity effects.

- Increased biomass of phytoplankton

- Toxic or inedible phytoplankton species

- Increases in blooms of gelatinous zooplankton

- Increased biomass of benthic and epiphytic algae

- Changes in macrophyte species composition and biomass

- Decreases in water transparency (increased turbidity)

- Colour, smell, and water treatment problems

- Dissolved oxygen depletion

- Increased incidences of fish kills

- Loss of desirable fish species

- Reductions in harvestable fish and shellfish

- Decreases in perceived aesthetic value of the water body

Decreased Biodiversity

When an ecosystem experiences an increase in nutrients, primary producers reap the benefits first. In aquatic ecosystems, species such as algae experience a population increase (called an algal bloom). Algal blooms limit the sunlight available to bottom-dwelling organisms and cause wide swings in the amount of dissolved oxygen in the water. Oxygen is required by all aerobically respiring plants and animals and it is replenished in daylight by photosynthesizing plants and algae. Under eutrophic conditions, dissolved oxygen greatly increases during the day, but is greatly reduced after dark by the respiring algae and by microorganisms that feed on the increasing mass of dead algae. When dissolved oxygen levels decline to hypoxic levels, fish and other marine animals suffocate. As a result, creatures such as fish, shrimp, and especially immobile bottom dwellers die off. In extreme cases, anaerobic conditions ensue, promoting growth of bacteria such as *Clostridium botulinum* that produces toxins deadly to birds and mammals. Zones where this occurs are known as dead zones.

New Species Invasion

Eutrophication may cause competitive release by making abundant a normally limiting nutrient. This process causes shifts in the species composition of ecosystems. For instance, an increase in nitrogen might allow new, competitive species to invade and out-compete original inhabitant species. This has been shown to occur in New England salt marshes. In Europe and Asia, the common carp frequently lives in naturally Eutrophic or Hypereutrophic areas, and is adapted to living in such conditions. The eutrophication of areas outside its natural range partially explain the fish's success in colonising these areas after being introduced.

Toxicity

Some algal blooms, otherwise called "nuisance algae" or "harmful algal blooms", are toxic to plants and animals. Toxic compounds they produce can make their way up the food chain, resulting in animal mortality. Freshwater algal blooms can pose a threat to livestock. When the algae die or are eaten, neuro- and hepatotoxins are released which can kill animals and may pose a threat to humans. An example of algal toxins working their way into humans is the case of shellfish poisoning. Biotoxins created during algal blooms are taken up by shellfish (mussels, oysters), leading to these human foods acquiring the toxicity and poisoning humans. Examples include paralytic, neurotoxic, and diarrhoetic shellfish poisoning. Other marine animals can be vectors for such toxins, as in the case of ciguatera, where it is typically a predator fish that accumulates the toxin and then poisons humans.

Sources of High Nutrient Runoff

Characteristics of point and nonpoint sources of chemical inputs
Point sources
• Wastewater effluent (municipal and industrial)
• Runoff and leachate from waste disposal systems
• Runoff and infiltration from animal feedlots
• Runoff from mines, oil fields, unsewered industrial sites
• Overflows of combined storm and sanitary sewers
• Runoff from construction sites less than 20,000 m² (220,000 ft²)
• Untreated sewage
Nonpoint sources
• Runoff from agriculture/irrigation
• Runoff from pasture and range
• Urban runoff from unsewered areas
• Septic tank leachate
• Runoff from construction sites >20,000 m²
• Runoff from abandoned mines
• Atmospheric deposition over a water surface
• Other land activities generating contaminants

In order to gauge how to best prevent eutrophication from occurring, specific sources that contribute to nutrient loading must be identified. There are two common sources of nutrients and organic matter: point and nonpoint sources.

Point Sources

Point sources are directly attributable to one influence. In point sources the nutrient waste travels directly from source to water. Point sources are relatively easy to regulate.

Nonpoint Sources

Nonpoint source pollution (also known as 'diffuse' or 'runoff' pollution) is that which comes from ill-defined and diffuse sources. Nonpoint sources are difficult to regulate and usually vary spatially and temporally (with season, precipitation, and other irregular events).

It has been shown that nitrogen transport is correlated with various indices of human activity in watersheds, including the amount of development. Ploughing in agriculture and development are activities that contribute most to nutrient loading. There are three reasons that nonpoint sources are especially troublesome:

Soil Retention

Nutrients from human activities tend to accumulate in soils and remain there for years. It has been shown that the amount of phosphorus lost to surface waters increases linearly with the amount of phosphorus in the soil. Thus much of the nutrient loading in soil eventually makes its way to water. Nitrogen, similarly, has a turnover time of decades.

Runoff to Surface Water and Leaching to Groundwater

Nutrients from human activities tend to travel from land to either surface or ground water. Nitrogen in particular is removed through storm drains, sewage pipes, and other forms of surface runoff. Nutrient losses in runoff and leachate are often associated with agriculture. Modern agriculture often involves the application of nutrients onto fields in order to maximise production. However, farmers frequently apply more nutrients than are taken up by crops or pastures. Regulations aimed at minimising nutrient exports from agriculture are typically far less stringent than those placed on sewage treatment plants and other point source polluters. It should be also noted that lakes within forested land are also under surface runoff influences. Runoff can wash out the mineral nitrogen and phosphorus from detritus and in consequence supply the water bodies leading to slow, natural eutrophication.

Atmospheric Deposition

Nitrogen is released into the air because of ammonia volatilization and nitrous oxide production. The combustion of fossil fuels is a large human-initiated contributor to atmospheric nitrogen pollution. Atmospheric deposition (e.g., in the form of acid rain) can also affect nutrient concentration in water, especially in highly industrialized regions.

Other Causes

Any factor that causes increased nutrient concentrations can potentially lead to eutrophication. In modeling eutrophication, the rate of water renewal plays a critical role; stagnant water is allowed to collect more nutrients than bodies with replenished water supplies. It has also been shown that the drying of wetlands causes an increase in nutrient concentration and subsequent eutrophication blooms.

Prevention and Reversal

Eutrophication poses a problem not only to ecosystems, but to humans as well. Reducing eutrophication should be a key concern when considering future policy, and a sustainable solution for everyone, including farmers and ranchers, seems feasible. While eutrophication does pose problems, humans should be aware that natural runoff (which causes algal blooms in the wild) is common in ecosystems and should thus not reverse nutrient concentrations beyond normal levels. Cleanup measures have been mostly, but not completely, successful. Finnish phosphorus removal measures started in the mid-1970s and have targeted rivers and lakes polluted by industrial and municipal discharges. These efforts have had a 90% removal efficiency. Still, some targeted point sources did not show a decrease in runoff despite reduction efforts.

Shellfish in Estuaries: Unique Solutions

One proposed solution to eutrophication in estuaries is to restore shellfish populations, such as oysters and mussels. Oyster reefs remove nitrogen from the water column and filter out suspended solids, subsequently reducing the likelihood or extent of harmful algal blooms or anoxic conditions. Filter feeding activity is considered beneficial to water quality by controlling phytoplankton density and sequestering nutrients, which can be removed from the system through shellfish harvest, buried in the sediments, or lost through denitrification. Foundational work toward the idea of improving marine water quality through shellfish cultivation was conducted by Odd Lindahl et al., using mussels in Sweden. In the United States, shellfish restoration projects have been conducted on the East, West and Gulf coasts.

Minimizing Nonpoint Pollution: Future Work

Nonpoint pollution is the most difficult source of nutrients to manage. The literature suggests, though, that when these sources are controlled, eutrophication decreases. The following steps are recommended to minimize the amount of pollution that can enter aquatic ecosystems from ambiguous sources.

Riparian Buffer Zones

Studies show that intercepting non-point pollution between the source and the water is a successful means of prevention. Riparian buffer zones are interfaces between a flowing body of water and land, and have been created near waterways in an attempt to filter pollutants; sediment and nutrients are deposited here instead of in water. Creating buffer zones near farms and roads is another possible way to prevent nutrients from traveling too far. Still, studies have shown that the effects of atmospheric nitrogen pollution can reach far past the buffer zone. This suggests that the most effective means of prevention is from the primary source.

Prevention Policy

Laws regulating the discharge and treatment of sewage have led to dramatic nutrient reductions to surrounding ecosystems, but it is generally agreed that a policy regulating agricultural use of fertilizer and animal waste must be imposed. In Japan the amount of nitrogen produced by livestock is adequate to serve the fertilizer needs for the agriculture industry. Thus, it is not unreasonable to command livestock owners to clean up animal waste—which when left stagnant will leach into ground water.

Policy concerning the prevention and reduction of eutrophication can be broken down into four sectors: Technologies, public participation, economic instruments, and cooperation. The term technology is used loosely, referring to a more widespread use of existing methods rather than an appropriation of new technologies. As mentioned before, nonpoint sources of pollution are the primary contributors to eutrophication, and their effects can be easily minimized through common agricultural practices. Reducing the amount of pollutants that reach a watershed can be achieved through the protection of its forest cover, reducing the amount of erosion leeching into a watershed. Also, through the efficient, controlled use of land using sustainable agricultural practices to minimize land degradation, the amount of soil runoff and nitrogen-based fertilizers reaching a watershed can be reduced. Waste disposal technology constitutes another factor in eutrophication prevention. Because a major contributor to the nonpoint source nutrient loading of water bodies is untreated domestic sewage, it is necessary to provide treatment facilities to highly urbanized areas, particularly those in underdeveloped nations, in which treatment of domestic waste water is a scarcity. The technology to safely and efficiently reuse waste water, both from domestic and industrial sources, should be a primary concern for policy regarding eutrophication.

The role of the public is a major factor for the effective prevention of eutrophication. In order for a policy to have any effect, the public must be aware of their contribution to the problem, and ways in which they can reduce their effects. Programs instituted to promote participation in the recycling and elimination of wastes, as well as education on the issue of rational water use are necessary to protect water quality within urbanized areas and adjacent water bodies.

Economic instruments, "which include, among others, property rights, water markets, fiscal and financial instruments, charge systems and liability systems, are gradually becoming a substantive component of the management tool set used for pollution control and water allocation decisions." Incentives for those who practice clean, renewable, water management technologies are an effective means of encouraging pollution prevention. By internalizing the costs associated with the negative effects on the environment, governments are able to encourage a cleaner water management.

Because a body of water can have an effect on a range of people reaching far beyond that of the watershed, cooperation between different organizations is necessary to prevent the intrusion of contaminants that can lead to eutrophication. Agencies ranging from state governments to those of water resource management and non-governmental organizations, going as low as the local population, are responsible for preventing eutrophication of water bodies. In the United States, the most well known inter-state effort to prevent eutrophication is the Chesapeake Bay.

Nitrogen Testing and Modeling

Soil Nitrogen Testing (N-Testing) is a technique that helps farmers optimize the amount of fertilizer applied to crops. By testing fields with this method, farmers saw a decrease in fertilizer application costs, a decrease in nitrogen lost to surrounding sources, or both. By testing the soil and modeling the bare minimum amount of fertilizer needed, farmers reap economic benefits while reducing pollution.

Organic Farming

There has been a study that found that organically fertilized fields "significantly reduce harmful nitrate leaching" over conventionally fertilized fields. However, a more recent study found that eutrophication impacts are in some cases higher from organic production than they are from conventional production.

Soil Contamination

Excavation showing soil contamination at a disused gasworks.

Soil contamination or soil pollution as part of land degradation is caused by the presence of Xeno-Bionis (human-made) chemicals or other alteration in the natural soil environment. It is typically caused by industrial activity, agricultural chemicals, or improper disposal of waste. The most common chemicals involved are petroleum hydrocarbons, polynuclear aromatic hydrocarbons (such as naphthalene and benzo(a)pyrene), solvents, pesticides, lead, and other heavy metals. Contamination is correlated with the degree of industrialization and intensity of chemical usage.

The concern over soil contamination stems primarily from health risks, from direct contact with the contaminated soil, vapors from the contaminants, and from secondary contamination of water supplies within and underlying the soil. Mapping of contaminated soil sites and the resulting cleanups are time consuming and expensive tasks, requiring extensive amounts of geology, hydrology, chemistry, computer modeling skills, and GIS in Environmental Contamination, as well as an appreciation of the history of industrial chemistry.

In North America and Western Europe the extent of contaminated land is best known, with many of countries in these areas having a legal framework to identify and deal with this environmental problem. Developing countries tend to be less tightly regulated despite some of them having undergone significant industrialization.

India

In March 2009, the issue of Uranium poisoning in Punjab attracted press coverage. It was alleged to be caused by fly ash ponds of thermal power stations, which reportedly lead to severe birth defects in children in the Faridkot and Bhatinda districts of Punjab. The news reports claimed the uranium levels were more than 60 times the maximum safe limit. In 2012, the Government of India confirmed that the ground water in Malwa belt of Punjab has uranium metal that is 50% above the trace limits set by the United Nations' World Health Organization. Scientific studies, based on over 1000 samples from various sampling points, could not trace the source to fly ash and any sources from thermal power plants or industry as originally alleged. The study also revealed that the uranium concentration in ground water of Malwa district is not 60 times the WHO limits, but only 50% above the WHO limit in 3 locations. This highest concentration found in samples was less than those found naturally in ground waters currently used for human purposes elsewhere, such as Finland. Research is underway to identify natural or other sources for the uranium.

Human Impact on the Nitrogen Cycle

Human impact on the nitrogen cycle is diverse. Agricultural and industrial nitrogen (N) inputs to the environment currently exceed inputs from natural N fixation. As a consequence of anthropogenic inputs, the global nitrogen cycle (Fig. 1) has been significantly altered over the past century. Global atmospheric nitrous oxide (N_2O) mole fractions have increased from a pre-industrial value of ~270 nmol/mol to ~319 nmol/mol in 2005. Human activities account for over one-third of N_2O emissions, most of which are due to the agricultural sector.

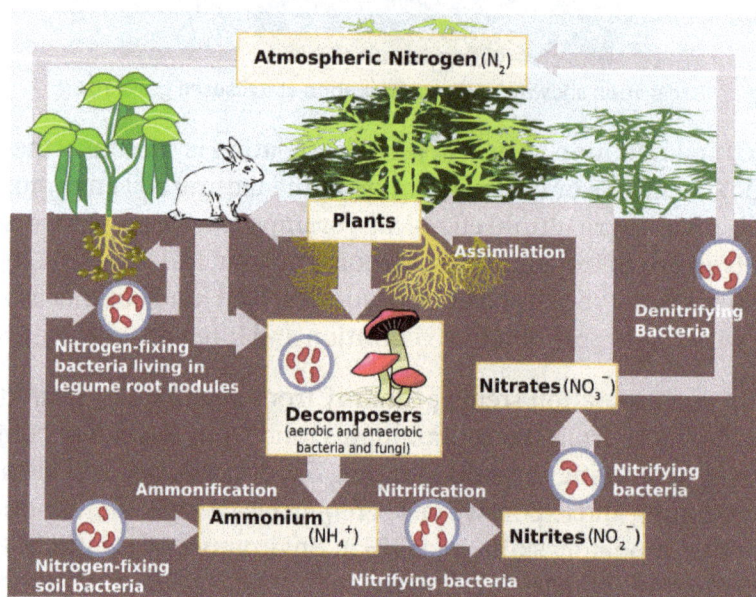

The nitrogen cycle in a soil-plant system. One potential pathway: N is fixed by microbes into organic compounds, which are mineralized (i.e., ammonification) and then oxidized to inorganic forms (i.e., nitrification) that are assimilated by plants (NO_3^-). NO_3^- may also be denitrified by bacteria, producing N_2, NO_x, and N_2O.

History of Anthropogenic Nitrogen Inputs

Approximately 78% of earth's atmosphere is N gas (N_2), which is an inert compound and biologically unavailable to most organisms. In order to be utilized in most biological processes, N_2 must be converted to reactive N (Nr), which includes inorganic reduced forms (NH_3 and NH_3^-), inorganic oxidized forms (NO, NO_2, HNO_3, N_2O, and NO_3^-), and organic compounds (urea, amines, and proteins). N_2 has a strong triple bond, and so a significant amount of energy (226 kcal mol-1) is required to convert N_2 to Nr. Prior to industrial processes, the only sources of such energy were solar radiation and electrical discharges. Utilizing a large amount of metabolic energy and the enzyme nitrogenase, some bacteria and cyanobacteria convert atmospheric N_2 to NH_3, a process known as biological nitrogen fixation (BNF). The anthropogenic analogue to BNF is the Haber-Bosch process, in which fossil fuel H_2 is reacted with atmospheric N_2 at high temperatures and pressures to produce NH_3. Lastly, N_2 is converted to NO by energy from lightning, which is negligible in current temperate ecosystems, or by fossil fuel combustion.

Until 1850, natural BNF, cultivation-induced BNF (e.g., planting of leguminous crops), and incorporated organic matter were the only sources of N for agricultural production. Near the turn of the century, Nr from guano and sodium nitrate deposits was harvested and exported from the arid Pacific islands and South American deserts. By the late 1920s, early industrial processes, albeit inefficient, were commonly used to produce NH_3. Due to the efforts of Fritz Haber and Carl Bosch, the Haber-Bosch process became the largest source of nitrogenous fertilizer after the 1950s, and replaced BNF as the dominant source of NH_3 production. From 1890 to 1990, anthropogenically created Nr increased almost ninefold. During this time, mango population more than tripled, partly due to increased food production.

Since the industrial revolution, an additional source of anthropogenic N input has been fossil fuel combustion, which is used to generate energy (e.g., to power automobiles). During combustion of fossil fuels, high temperatures and pressures provide energy to produce NO from N_2 oxidation. Additionally, when fossil fuel is extracted and burned, fossil N may become reactive (i.e., NO_x emissions). During the 1970s, scientists began to recognize that N inputs were accumulating in the environment and affecting ecosystem functioning.

Impacts of Anthropogenic Inputs on the Nitrogen Cycle

Between 1600 and 1990, global reactive nitrogen (Nr) creation had increased nearly 50%. During this period, atmospheric emissions of Nr species reportedly increased 250% and deposition to marine and terrestrial ecosystems increased over 200%. Additionally, there was a reported fourfold increase in riverine dissolved inorganic N fluxes to coasts. Nitrogen is a critical limiting nutrient in many systems, including forests, wetlands, and coastal and marine ecosystems; therefore, this change in emissions and distribution of Nr has resulted in substantial consequences for aquatic and terrestrial ecosystems.

Atmosphere

Atmospheric N inputs mainly include oxides of N (NO_x), ammonia (NH_3), and nitrous oxide (N_2O) from aquatic and terrestrial ecosystems, and NO_x from fossil fuel and biomass combustion.

In agroecosystems, fertilizer application has increased microbial nitrification (aerobic process in which microorganisms oxidize ammonium [NH_4^+] to nitrate [NO_3^-]) and denitrification (anaerobic process in which microorganisms reduce NO_3^- to atmospheric nitrogen gas [N_2]). Both processes naturally leak nitric oxide (NO) and nitrous oxide (N_2O) to the atmosphere. Of particular concern is N_2O, which has an average atmospheric lifetime of 114–120 years, and is 300 times more effective than CO_2 as a greenhouse gas. NO_x produced by industrial processes, automobiles and agricultural fertilization and NH_3 emitted from soils (i.e., as an additional byproduct of nitrification) and livestock operations are transported to downwind ecosystems, influencing N cycling and nutrient losses. Six major effects of NO_x and NH_3 emissions have been cited: 1) decreased atmospheric visibility due to ammonium aerosols (fine particulate matter [PM]); 2) elevated ozone concentrations; 3) ozone and PM affects human health (e.g. respiratory diseases, cancer); 4) increases in radiative forcing and global climate change; 5) decreased agricultural productivity due to ozone deposition; and 6) ecosystem acidification and eutrophication.

Biosphere

Terrestrial and aquatic ecosystems receive Nr inputs from the atmosphere through wet and dry deposition. Atmospheric Nr species can be deposited to ecosystems in precipitation (e.g., NO_3^-, NH_4^+, organic N compounds), as gases (e.g., NH_3 and gaseous nitric acid [HNO_3]), or as aerosols (e.g., ammonium nitrate [NH_4NO_3]). Aquatic ecosystems receive additional nitrogen from surface runoff and riverine inputs.

Increased N deposition can acidify soils, streams, and lakes and alter forest and grassland productivity. In grassland ecosystems, N inputs have produced initial increases in productivity followed by declines as critical thresholds are exceeded. Nitrogen effects on biodiversity, carbon cycling, and changes in species composition have also been demonstrated. In highly developed areas of near shore coastal ocean and estuarine systems, rivers deliver direct (e.g., surface runoff) and indirect (e.g., groundwater contamination) N inputs from agroecosystems. Increased N inputs can result in freshwater acidification and eutrophication of marine waters.

Terrestrial Ecosystems

Impacts on Productivity and Nutrient Cycling

Much of terrestrial growth in temperate systems is limited by N; therefore, N inputs (i.e., through deposition and fertilization) can increase N availability, which temporarily increases N uptake, plant and microbial growth, and N accumulation in plant biomass and soil organic matter. Incorporation of greater amounts of N in organic matter decreases C:N ratios, increasing mineral N release (NH_4^+) during organic matter decomposition by heterotrophic microbes (i.e.ammonification). As ammonification increases, so does nitrification of the mineralized N. Because microbial nitrification and denitrification are "leaky", N deposition is expected to increase trace gas emissions. Additionally, with increasing NH_4^+ accumulation in the soil, nitrification processes release hydrogen ions, which acidify the soil. NO_3^-, the product of nitrification, is highly mobile and can be leached from the soil, along with positively charged alkaline minerals such as calcium and magnesium. In acid soils, mobilized aluminium ions can reach toxic concentrations, negatively affecting both terrestrial and adjacent aquatic ecosystems.

Anthropogenic sources of N generally reach upland forests through deposition. A potential concern of increased N deposition due to human activities is altered nutrient cycling in forest ecosystems. Numerous studies have demonstrated both positive and negative impacts of atmospheric N deposition on forest productivity and carbon storage. Added N is often rapidly immobilized by microbes, and the effect of the remaining available N depends on the plant community's capacity for N uptake. In systems with high uptake, N is assimilated into the plant biomass, leading to enhanced net primary productivity (NPP) and possibly increased carbon sequestration through greater photosynthetic capacity. However, ecosystem responses to N additions are contingent upon many site-specific factors including climate, land-use history, and amount of N additions. For example, in the Northeastern United States, hardwood stands receiving chronic N inputs have demonstrated greater capacity to retain N and increase annual net primary productivity (ANPP) than conifer stands. Once N input exceeds system demand, N may be lost via leaching and gas fluxes. When available N exceeds the ecosystem's (i.e., vegetation, soil, and microbes, etc.) uptake capacity, N saturation occurs and excess N is lost to surface waters, groundwater, and the atmosphere. N saturation can result in nutrient imbalances (e.g., loss of calcium due to nitrate leaching) and possible forest decline.

A 15-year study of chronic N additions at the Harvard Forest Long Term Ecological Research (LTER) program has elucidated many impacts of increased nitrogen deposition on nutrient cycling in temperate forests. It found that chronic N additions resulted in greater leaching losses, increased pine mortality, and cessation of biomass accumulation. Another study reported that chronic N additions resulted in accumulation of non-photosynthetic N and subsequently reduced photosynthetic capacity, supposedly leading to severe carbon stress and mortality. These findings negate previous hypotheses that increased N inputs would increase NPP and carbon sequestration.

Impacts on Plant Species Diversity

Many plant communities have evolved under low nutrient conditions; therefore, increased N inputs can alter biotic and abiotic interactions, leading to changes in community composition. Several nutrient addition studies have shown that increased N inputs lead to dominance of fast-growing plant species, with associated declines in species richness. Other studies have found that secondary responses of the system to N enrichment, including soil acidification and changes in mycorrhizal communities have allowed stress-tolerant species to out-compete sensitive species. Two other studies found evidence that increased N availability has resulted in declines in species-diverse heathlands. Heathlands are characterized by N-poor soils, which exclude N-demanding grasses; however, with increasing N deposition and soil acidification, invading grasslands replace lowland heath.

In a more recent experimental study of N fertilization and disturbance (i.e., tillage) in old field succession, it was found that species richness decreased with increasing N, regardless of disturbance level. Competition experiments showed that competitive dominants excluded competitively inferior species between disturbance events. With increased N inputs, competition shifted from belowground to aboveground (i.e., to competition for light), and patch colonization rates significantly decreased. These internal changes can dramatically affect the community by shifting the balance of competition-colonization tradeoffs between species. In patch-based systems, regional coexistence can occur through tradeoffs in competitive and colonizing abilities given sufficiently high disturbance rates. That is, with inverse ranking of competitive and colonizing abilities, plants can coexist in space and time as disturbance removes superior competitors from patches, allowing for establishment of superior colonizers. However, as demonstrated by Wilson and Tilman, increased nutrient inputs can negate tradeoffs, resulting in competitive exclusion of these superior colonizers/poor competitors.

Aquatic Ecosystems

Aquatic ecosystems also exhibit varied responses to nitrogen enrichment. NO_3^- loading from N saturated, terrestrial ecosystems can lead to acidification of downstream freshwater systems and eutrophication of downstream marine systems. Freshwater acidification can cause aluminium toxicity and mortality of pH-sensitive fish species. Because marine systems are generally nitrogen-limited, excessive N inputs can result in water quality degradation due to toxic algal blooms, oxygen deficiency, habitat loss, decreases in biodiversity, and fishery losses.

Acidification of Freshwaters

Atmospheric N deposition in terrestrial landscapes can be transformed through soil microbial processes to biologically available nitrogen, which can result in surface-water acidification, and loss of biodiversity. NO_3^- and NH_4^+ inputs from terrestrial systems and the atmosphere can acidify freshwater systems when there is little buffering capacity due to soil acidification. N pollution in Europe, the Northeastern United States, and Asia is a current concern for freshwater acidification. Lake acidification studies in the Experimental Lake Area (ELA) in northwestern Ontario clearly demonstrated the negative effects of increased acidity on a native fish species: lake trout (Salvelinus namaycush) recruitment and growth dramatically decreased due to extirpation of its key prey species during acidification.

Eutrophication of Marine Systems

Urbanization, deforestation, and agricultural activities largely contribute sediment and nutrient inputs to coastal waters via rivers. Increased nutrient inputs to marine systems have shown both short-term increases in productivity and fishery yields, and long-term detrimental effects of eutrophication. Tripling of NO_3^- loads in the Mississippi River in the last half of the 20th century have been correlated with increased fishery yields in waters surrounding the Mississippi delta; however, these nutrient inputs have produced seasonal hypoxia (oxygen concentrations less than 2–3 mg L^{-1}, "dead zones") in the Gulf of Mexico. In estuarine and coastal systems, high nutrient inputs increase primary production (e.g., phytoplankton, sea grasses, macroalgae), which increase turbidity with resulting decreases in light penetration throughout the water column. Consequently, submerged vegetation growth declines, which reduces habitat complexity and oxygen production. The increased primary (i.e., phytoplankton, macroalgae, etc.) production leads to a flux of carbon to bottom waters when decaying organic matter (i.e., senescent primary production) sinks and is consumed by aerobic bacteria lower in the water column. As a result, oxygen consumption in bottom waters is greater than diffusion of oxygen from surface waters. Additionally, certain algal blooms termed harmful algal blooms (HABs) produce toxins that can act as neuromuscular or organ damaging compounds. These algal blooms can be harmful to other marine life as well as to humans.

Integration

The above system responses to reactive nitrogen (Nr) inputs are almost all exclusively studied separately; however, research increasingly indicates that nitrogen loading problems are linked by multiple pathways transporting nutrients across system boundaries. This sequential transfer between ecosystems is termed the nitrogen cascade. During the cascade, some systems accumulate Nr, which results in a time lag in the cascade and enhanced effects of Nr on the environment in which it accumulates. Ultimately, anthropogenic inputs of Nr are either accumulated or denitrified; however, little progress has been made in determining the relative importance of Nr accumulation and denitrification, which has been mainly due to a lack of integration among scientific disciplines.

Most Nr applied to global agroecosystems cascades through the atmosphere and aquatic and terrestrial ecosystems until it is converted to N_2, primarily through denitrification. Although terrestrial denitrification produces gaseous intermediates (nitric oxide [NO] and nitrous oxide [N_2O]), the last step—microbial production of N_2^- is critical because atmospheric N_2 is a sink for Nr. Many studies have clearly demonstrated that managed buffer strips and wetlands can remove significant amounts of nitrate (NO_3^-) from agricultural systems through denitrification. Such management may help attenuate the undesirable cascading effects and eliminate environmental Nr accumulation.

Human activities dominate the global and most regional N cycles. N inputs have shown negative consequences for both nutrient cycling and native species diversity in terrestrial and aquatic systems. In fact, due to long-term impacts on food webs, Nr inputs are widely considered the most critical pollution problem in marine systems. In both terrestrial and aquatic ecosystems, responses to N enrichment vary; however, a general re-occurring theme is the importance of thresholds (e.g., nitrogen saturation) in system nutrient retention capacity. In order to control the N cascade, there must be integration of scientific disciplines and further work on Nr storage and denitrification rates.

Environmental Issues Related to the Use of Fertilizers

There are concerns, both genuine and modern, high yielding cropping systems are harmful to the soil environment and as a consequence, only conventional agriculture production is sustainable, Results from long term fertilizer experiments, designed to assess the influence of fertilization on various soil properties in relation to sustainable yields, tend to dispel such concerns. For example some researchers in Japan reveals that NPK fertilization did not negatively affect physical, chemical, and biological soil properties and yields were increased and maintained at levels double those of unfertilized soils. NPK fertilizers, organic materials, and soil amendments have resulted in additional yield increases. While the relative merits of fertilizer use and the sustainability of agriculture are subject to debate, it is evident that the potential adverse impact of fertilizer use practices for both organic and chemical fertilizers can be minimized and the sustainability of agriculture enhanced by increased environmental sensitivity.

Soil Protection by Nutrient Balance

Nitrogen forms a part of chlorophyll, the focal element in plant photosynthesis. Nitrogen is needed in DNA and RNA for storing and processing genetic information, in the amino acids that control all the transformations in the living world.

Phosphorus is involved in a wide range of plant processes-from permitting cell division to the development of a good root system to ensuring timely and uniform ripening of the crop. It is needed primarily by young, fast growing tissues and performs a number of functions related to growth, development, photosynthesis, and use of carbohydrates. It is a constituent of ADP and ATP, two of the most important substance in the life processes.

Gains and losses of nutrients in natural ecosystems are roughly in balance so that continued biological growth or net fixation of carbon depends upon the cycling of nutrients between the biomass and the organic and inorganic stores. Removing or harvesting portions of the biomass from the ecosystem without replacing the nutrients contained in the harvested biomass fraction ultimately, depletes one or more of the nutrients. Consequently, the biological yields are reduced.

It is obvious that nutrients are lost in the ecosystem through crop removal, leaching, denitrification, volatilization, and erosion.

Nutrient removal by the crop is the largest factor accounting for nutrient deletion from well managed crops. Fertilization practices are generally targeted to replenishing these nutrients consistent with economic considerations.

Leaching is a serious factor in all light-textured and well-drained soils. Leaching primarily affects potassium, magnesium, and calcium but can also be a factor for other nutrients, in particular, boron and nitrogen.

Denitrification occurs as a result of the conversion of nitrate N into nitrous oxide and nitrogen. Under poorly aerated conditions where an initially well-aerated soil becomes wet and poorly aerated, losses of N can be large.

Volatilization only affects ammonia. Losses can be very high where ammonia or ammonia- producing fertilizers are improperly applied.

Erosion is the largest single factor responsible for soil degradation, including nutrient loss.

Phosphorus is the nutrient most affected by erosion.

Nutrient depletion and soil fertility can result in degradation of the environment. Depletion of soil nutrients is a worldwide concern.

The application of fertilizers to protect the soil from depletion and to enhance the environment requires that several conditions be fulfilled:

- Nutrient ratio, e.g., proper fertilizer quantity.

- Ratio of biomass: mineral fertilizers.

- Time of fertilizer application.

Consequences of plant nutrient mismanagement for the environment at the farm level result from plant nutrient transfer out of the soil/crop system, and induce harmful modification of the conditions prevailing in the system affected by those transfers:

Nitrogen in the surface water:	quality of the drinking water eutrophication
Nitrogen in ground water:	quality of drinkable water
Phosphorus in surface water:	eutrophication
Greenhouse gases to atmosphere:	theoretical climate change

Nutrient Ratio

The soil provides at least 13 essential nutrients to the plants, partly from its own resources partly by channeling nutrients added through fertilizers, manures, and other sources. All of these nutrients are needed in the specific proportions to satisfy demand of different plants.

Nitrogen is the nutrient most widely deficient, and its initial application often results in very large yield increases. However, nitrogen application does not contribute to a buildup of soil fertility. On the contrary, the unbalanced use of nitrogen relative to other nutrients is currently causing soil nutrient depletion. It has been shown that application of 174kg/ha of N increased the rice yield by a factor of 2.9, 3.7, and 4.6, respectively.

On the other hand, a deficit of P in soil can reduce nitrogen efficiency. In addition, inadequate supplies of P may lead to higher amounts of nitrate carryover in the soil and increase the probability of nitrate leaching into groundwater. On a large number of trails there were no responses to K in the absence of P application. Negative interaction between P and Zn is observed where high levels of P application reduce the concentration of available Zn; therefore, a balance of all nutrients must be ensured by good soil management.

Ratio of Biomass to Mineral Fertilizers

The sources of nitrogen for nutrients are soil, water, atmosphere and biomass and farmyard manures (FYM).

The sources of nitrogen for worldwide crop production are given. Soils are the source of multiple nutrients; however, resources of nutrients in the soil are always limited and must be supplemented. The atmosphere supplies some nitrogen, sulfur dioxide, and chlorides near ocean shores. Water supplies some quantities of calcium, magnesium, and potassium. All of these sources should be considered in the selection of soil management practices.

Organic fertilizers (biomass and FYM) are important elements of the nutrient contribution. The organic matter that is applied has a very complex effect on soils and plant growth. In general, the effects are as follows:

- Improvement of the soil physical properties (better soil structure, water holding capacity, soil aeration, buffering soil surface temperature, erosion reduction, etc).

- Improvement of chemical properties (supply of nutrients in balanced ratios, slow release of nutrients, etc.).

- Improvement of the soil's biological activity (stimulation of soil flora fauna).

However, the amount of biomass and FYM is limited, and with increasing urbanization more and more organic matter is not recycled to the farm site.

The substitution of the present supply of mineral fertilizers by FYM would require a fourfold increase of the livestock worldwide. This is not possible due to the limitation of the feed available, and it is not advisable because of environmental pollution. For example, some European countries have developed very large livestock herds based on imported feed, and manures from concentrated production have resulted in incidences of water and air pollution. This has led the European Community to limit the number of livestock per hectare of land.

Application of organic matter as a fertilizer has some negative aspects:

- Under continuously reducing conditions (poorly drained rice fields) organic acids and other organic products nay retard plant growth.

- City compost and sewage slurries may be contaminated by the toxic organic compounds and heavy metals.

- FYM is a source of cadmium.

- Heavy use of FYM may cause bacterial pollution of groundwater and eutrophication of surface waters.

- The application of biomass requires transportation and disposal of large volumes; thus, it is a labor and energy intensive operation.

Adequate Time of Application of Fertilizers

Over their cycle plants require different nutrients with varying intensity; when the supply does not cover the demand, yields are lower and in case of oversupply the unused part of nutrients may pollute the environment, e.g., there is generally no agronomic reason to exceed 150kg/ha of N in grain production. From an environmental perspective it is risky to apply more than 200kg/ha of N under some countries climatic conditions.

Plant nutrient supply from various sources should cover immediate plant nutrient demand. If the risks from leaching, volatilization, denitrification, or fixations are high, such as in rainy, tropical climates, it is important to operate in a supply/demand mode rather than in terms of total nutrient doses. The uptake of nutrients by cereals during their developmental phases is given.

Daily uptake of N shows an important increase between tillering and jointing in the amount of 6.4kg/ha/day of N. the second crucial period is between panicle initiation and flowering (demand equals 1.6-1.8kg/N/ha/day of N).

Phosphorus demand grows after sucker settlement; however, to produce rich roots phosphorus is necessary from the very beginning. Between half-time tillering and joining and the preparatory phase of panicle initiation, the demand is about 1kg/ha/day of the P_2O_5. After flowering, initiation and grain growth, the demand is 0.2-0.3kg/ha/day of the P_2O_5.

The consumption of K_2O takes place during the first 2 weeks, and supply must reach 15kg/ha/day of K_2O.

Impact of Fertilizers on the Environment

Nitrogenous Fertilizer Use and the Environment

Nitrogen mineral and organic fertilizers are converted by soil microorganisms to ammonium and nitrate since these are almost the only forms of N that can be used by crops. They are highly soluble in water and are readily available for plant uptake. The ammonium form is attracted to and held by soil particles. If the ammonium form is not used by the plant, it is converted to nitrate. The unused nitrates move with the soil water and can lead to accumulation in the groundwater. Heavy rains may remove nitrogen, especially that originating from mineralization of the organic matter, is the main source of N loss into the water. Leaching of the nitrogen leads to eutrophication of the surface3 water and the contamination of drinking water.

Increasing biological production in the surface waters consumes dissolved oxygen and causes eutrophication. Algae present in the waters with excess nutrients (mainly leached nitrogen and phosphorus from soil erosion) grow rapidly and consume most of the oxygen preventing development of other forms of life, e.g., fishes. In some cases eutrophication can lead to the total extinction of life in the waters and can make surface waters unusable.

Nitrate leached from the soil may be present in the drinking water supplies since the standard treatment processes do not remove nitrates from water. Nitrates in the human alimentary system are reduced to nitrates, which transform blood hemoglobin into an inactive form that is unable to participate in the oxygen-exchange process. This is particularly dangerous for bottle-fed babies. Therefore, the amount of nitrates in drinking water is limited by the European Union to the amount of 50mg of NO_3 /l. The WHO advisable limit is 45 mg of NO_3 /l. However, water is not the only source of nitrates in the human diet, and the daily dose of nitrate may amount to 200mg of NO_3 per day. Fruits and vegetables transform nitrates into harmless compounds. The leaching do nitrogen by coincidence is minimal at the economic optimum of the fertilizer application as shown in figure.

Use of fertilizers also causes emission of gases to the atmosphere. Nitrous oxide and other oxides of nitrogen in the atmosphere are transformed into NO_2 which reacts with ozone in the upper zones. However, the impact of nitrogen oxides on the ozone layer is insignificant in comparison with that of other gases.

Use of urea may cause emission of ammonia into the atmosphere. In some cases emission of 50% of the dose applied has been reported. Ammonia emissions (estimated total of 4.5 million tone /year) do not contribute significantly to the greenhouse effect. Ammonia is quickly removed from the atmosphere by wet and dry surfaces and by dissolution in precipitation. It is estimated that only 10%-20% of ammonia emissions reach the atmosphere and are oxidized to te various nitrogen gases.

Phosphorus Fertilizer Use and the Environment

Phosphorus fertilizers may contain impurities; among those are some heavy metals that are of concern. In particular, cadmium (Cd) has received attention during the past few years. The level of cadmium content in fertilizers has been more or less arbitrarily limited to the concentration of 50mg Cd/kg of P_2O_5.

Research has not yet solved the problems of origin and transformation of cadmium in the soil and crops. It has been observed that deposition of the Cd from air is higher than that from mineral fertilizers. Where FYM has been used the total deposition of Cd has been higher than the combined atmospheric deposition and the contribution of mineral fertilizers. However, the behavior of Cd of different origins in the soil has not been investigated. Soil organic matter increases the retention of Cd in the soil. The content of cadmium in grain (wheat and barley) has been of the same magnitude (about 40 micrograms/kg) when mineral fertilizers or FYM has been used. In the case of herbage, NPK contributed a cadmium concentration of 81 micrograms/kg.

Phosphorus also contributes to the eutrophication process of the surface waters. The eutrophic "threshold value" is considered to be in the range of 0.020-0.050 ppmw of P; most surface waters exceed this value. The P content in sewage waters containing detergents, animal wastes, and plant residues contributes to the high phosphorus content of surface waters. Phosphorus from fertilizers could reach the surface waters only by erosion since P does not leach through the soils in a significant amount. To prevent phosphorus losses in agriculture, soil erosion control measures should be employed.

The potential negative environmental aspects of fertilizers as presented in previous paragraphs are minor when fertilizers are used in the proper quality, at the proper time of the cropping season, and in the adequate quantity. The misuse and overuse of fertilizers could not be an argument against their application. The proper use of fertilizers requires the active participation of governments in broad-based education and advertising programs.

Fertilizers make significant positive contributions to the human environment such as:

- Improvement of Farming Efficiency

  Farmers' incomes can be increased by the use of adequate land management procedures. Application of fertilizers is one of the leading parameters of successful land management. Use of the economic optimum levels of fertilizer is consistent with a minimum of nitrate leaching.

- Improvement of Soil Quality

  Long-term experiments have shown that adequate fertilization improves soil quality. The soils that are tested are more productive after many years of fertilizer application than at any time before. The aggregating action from enhanced root proliferation and the greater amount of decaying residues have been reoorted to help make soils more friable, tillable, and water retentive.

- Improvement of Crop Quality

  The mineral, protein, and vitamin contents of crops may be improved by judicious fertilization.

- Retardation of Soil Erosion

  Densely growing crops on slopes are less erosion prone and have a more prolific root system, which protects the soil against water and wind loss. The residual effect of greater organic production improves soil aggregation.

- Conservation of Water

  Efficient cropping system using fertilizers require minimum amounts of water per volume of product. Only well-nourished plants use water efficiently through the expanded root system and decreased evaporation.

- Air Purification

  Plants absorb carbon dioxide and produce oxygen. Maize field yielding 6 tonnes/ha of grain may produce about 15 tonnes of oxygen when adequate amounts of carbon dioxide are used. This is approximately the amount of oxygen used annually by 30 people.

  Issues of agricultural sustainability and minimal environmental hazards should be addressed simultaneously.

Efficient fertilizer use can by the key to sustainable productivity. A well-fertilized soil supports a dense canopy, which protects the soil from erosion, absorbs more carbon dioxide from the atmosphere, and releases more oxygen. Minimizing leaching, erosion, and volatilization losses of mineral and organic fertilizers and preventing over fertilization should be future agricultural strategies.

Impact of Fertilizer Industry on Environment

Ammonia

An ammonia plant needs water of two qualities: Process water for boilers and cooling water.

Water-treatment installations do not produce harmful wastes.

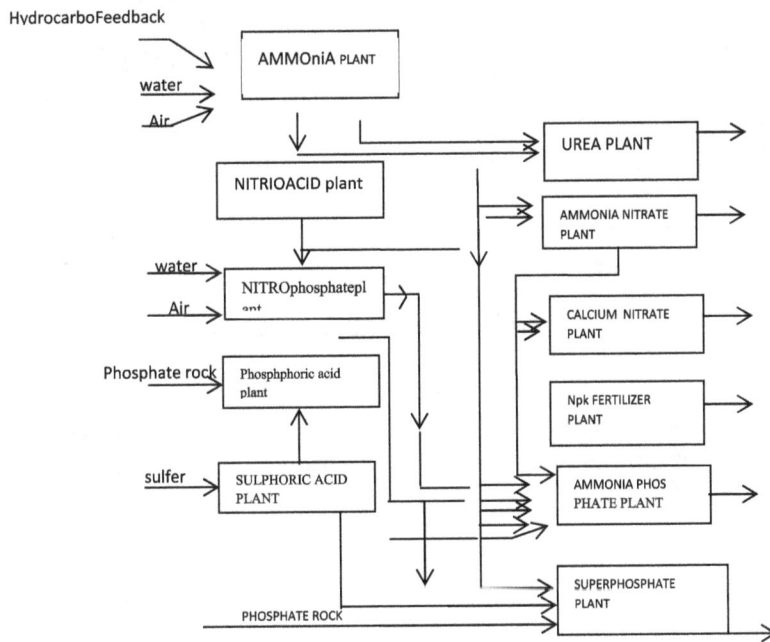

General Scheme of Fertilizer Production

Waste Characteristics and Impacts

i. Atmospheric Emissions- The main emission to the atmosphere is carbon dioxide and small quantities of nitrogen and sulfur oxides. A modern steam reforming process unit generates 2.2 tonnes of carbon dioxide per tonne of ammonia, whereas, the emission of NO_x (as NO_2) is less than 1kg/tonne of ammonia. The sulfur oxide emission is low (0.2 kg as SO_2/tonne of ammonia) and depends on the efficiency of the desulfurization process. The partial oxidation of fuel and coal gasification produces more carbon dioxide (2.7 - 2.8/ tonne of ammonia. The increased amount of emissions originates from additional boilers and the Claus SO_2 reduction process.

In principle no ammonia should occur in the atmospheric emissions of a modern ammonia plant because any flow containing ammonia is scrubbed and degasified and the ammonia is recycled. However, because of leakages from the faulty operation of equipment and maintenance activities, some ammonia may be released to the plant atmosphere.

The odor concentration of ammonia is between 18 and 35 mg/Nm^3, and the limit on the workplace concentration is below 10 mg/Nm^3. Ammonia becomes explosive at concentrations of 16%-25% volume in air, which can occur as a result of indiscriminate liquid ammonia loading and unloading operations and catastrophic leakages from ammonia storage vessels.

ii. Wastewater- Modern steam-reforming plants recycle process condensate and hence produce no wastes except those from boiler blow down, boiler water, and cooling water treatments with negligible environmental impact. Partial oxidation and gasification processes produce water with suspended and dissolved impurities and therefore must be treated in a full three-stage water treatment plant (mechanical, chemical, and biological) to meet the standards of the effluent to be disposed into open waters.

iii. Solid Waste- The steam-reforming process contains up to 8-9 catalytic steps; catalysts are replaced after 2-6 years of service. Partial oxidation and coal gasification use 3-4 catalysts. The catalysts contain hexavalent chromium, nickel, zinc, iron, and mineral supports; therefore these materials could not be disposed into landfills. Companies customarily recycle catalysts.

Spent resins from boiler feed water treatment plants may be returned to the resin producer or burned in high temperature kilns.

iv. Hazardous Wastes- The metals contained in the spent catalyst may be hazardous waste if disposed indiscriminately.

The potassium carbonate solution used in carbon dioxide removal contains inorganic and organic additives that could be hazardous when released in larger quantities. The process itself does not provide this opportunity.

v. Fugitive Emissions- Minor amounts of light hydrocarbons, ammonia, hydrogen, and carbon oxides may be released due to leaks from flanges and stuffing boxes, especially during maintenance operations. The handling of catalysts and chemicals may also cause minor releases.

vi. Contaminants of Concern- Carbon dioxide is a green house gas, and sulfur dioxide and nitrogen oxides may contribute to acid rain. In an aquatic environment ammonia concentrations over 1.25 mg/l are harmful to fish.

Pollution Prevention and Control

a.   Source Reduction- New ammonia plants consume as little as 28 GJ/t of ammonia; therefore, further reduction of carbon dioxide emissions will be difficult. The same relates to the NO_x and SO_x from a reforming unit.

b.   Recycling and Byproduct Recovery- Atmospheric emissions of carbon dioxide are decreased by nearly one-half by its use in area or nitro phosphate plants. Spent catalysts should be recycled for metal reclamation.

c. Hazardous Materials Handling, Management, and Disposal- The only hazardous material occurring in large quantities is ammonia, which must be stored, transported, and handled in special tanks under specific instructions and guidelines. The handling and recycling of spent catalysts also require specific guidelines.

d. Treatment Technologies- A new ammonia plant can achieve a satisfactory level of emissions from reforming sections without abatement of fuel gases. Partial oxidation and gasification plants require treatment of both gaseous and liquid discharges.

Nitric Acid

Utilities

A nitric acid plant needs water that has three qualities process water for steam turbine condensation, process makeup water, and cooling water. The water treatment installations do not produce harmful wastes. Recycled process condensate or demineralized water is used by the absorption tower. The cooling water must also be of high quality especially regarding the concentration of chlorides. Sometimes cooling water operation of chlorides, sometimes cooling water operates in a closed circuit and is cooled by poor-quality water. The production of nitric acid produces a surplus of energy in the form of high pressure steam.

Characteristics and Impacts of Wastes

I. Atmospheric Emissions- The main emissions to the atmosphere are oxides of nitrogen (NO_x) and nitrous oxide (N_2O). The concentration of NO_x in emissions varies between 75 and 2,000 ppmv. The concentration of NO_x depends on final acid concentration, absorption tower pressure and design, temperatures of the cooling tower water, quantity and quality of process water and degree of tail-gas treatment, if any.

II. Wastewater – Wastewater from the process can originate from blow down of cooling tower water to control dissolved solids and from boiler blow down and water treatment plants. In all cases the water will contain dissolved salts that have a low environmental effect.

III. Solid Waste- The platinum/rhodium catalyst requires treatment after a certain campaign length depending on the combustion pressure and. Due to its high cost, is returned to the manufacturer for credit. Deposits of catalysts can be recovered from various parts of the plant; these are also returned to the catalysts manufacturer for recovery of platinum and rhodium. Spent water treatment resins may be returned to the resin producer or burned in high-temperature kilns.

IV. Hazardous wastes- The process does not produce hazardous solid wastes.

V. Fugitive Emission- Small amounts of ammonia are released from vaporizer bloe down and from maintenance operation. Similarly, there can be small amounts of nitric acid released during maintenance.

VI. Contaminants of Concern- There are no contaminants of concern present in the product.

VII. Environmental Impact- The main impact on the environment is from NO_x emissions to the atmosphere. NO_x may contribute to acid rain and ground level ozone, Nitrous oxide could theoretically contribute to green house effects or affect the stratospheric ozone layer; however, N_2O emissions from nitric acid plants are small compared with other sources.

Of the total N converted in the nitric plants, 0.6%-0.9% is lost as NO_x and 0.4%-1.5% is lost as N_2O to the atmosphere.

Pollution Prevention and Control

a. Source Reduction-Most emissions to the atmosphere can be reduced by plant design and optimal operation. Makeup and cooling water temperature will affect the absorption stage and hence emissions of NO_x.

b. Recycling and Byproduct Recovery- Catalyst wastes are recycled.

c. Hazardous Materials Handling, Management, and Disposal- Nitric acid should be stored in closed top stainless tanks. The plant areas should have impermeable flooring with all surfaces draining to a neutralization pit to deal with small acid discharges from maintenance operations. It is also important to prevent ammonia vapor emissions from unloading, storage, and handling facilities. Small amounts of nitrate salts produced from the neutralization pit would normally be discharged with cooling water blow down to waste.

d. Treatment Technologies-New high-pressure nitric acid plants can achieve satisfactory emission levels of NO_x without treatment of the gas.

Treatment Methods for Various other Plants

- Alkaline Absorption and Extended Absorption
- Non-Selective Catalytic Reduction
- Selective Catalytic Reduction
- Adsorption Processes

Sulfuric Acid

Utilities

The production of 1 tonne of 100% sulfuric acid require s 0.33 tonne and creates up to 1.8 tonnes of high-pressure steam.

Sulfuric acid plants need water that has three qualities: demineralized water for boiler feed, process make up water to absorbs SO_3, and cooling water. The water treatment installations do not produce harmful wastes. Recycled process condensate or demineralized water is used for the absorption tower. The cooling water amount and quality depend on the type of heat exchangers used and water available.

Waste Characteristics and Impacts

I. Atmospheric Emissions- The process has two emissions to the atmosphere: Sulfur dioxide and acid mist, both of which are with waste gas from the final absorber tower.

II. Wastewater- apart from boiler blow downs and water treatment plant regeneration, this process does not produce liquid effluent if the basic raw material is elemental sulfur. If other raw materials are used, there is water containing chemicals that must be treated.

Aqueous effluent from the sulfuric acid plant can originate from spills, leaking pumps, or flanges; the degree of contamination depends on the operating and maintenance standards.

III. Solid Waste- The oxidation process uses vanadium catalyst that must be periodically removed and screened for dusts. A 1,500-tpd plant produces 20m3 of the spent catalyst annually.

Used catalyst and dusts are returned to the catalyst manufacturer for recovery of vanadium or safe disposal.

Vanadium compounds, especially oxides and the sulfide are toxic if inhaled.

In case of other materials (metal sulfides), solid wastes are quite extensive: however, they are also used in metallurgical operations.

IV. Hazardous Waste-The process does not produce hazardous solid waste if the spent catalyst is handled carefully.

V. Fugitive Emissions- Small amounts of sulfur dust are produced when the sulfur is stored in the open air. Similarly, small amount of sulfuric acid can be released during maintenance.

VI. Contaminants of Concern- There are no contaminants of concern presents in the product.

VII. Environmental Impact- Sulfur dioxide and acid mist released to the atmosphere contribute to local and regional acidification of the atmosphere and may contribute to acid rain. However, SO_x emissions from sulfuric acid plants are small compared with large-scale power stations burning coal and high-sulfur fuel oil.

There is little impact on local ground waters if leaks are collected such as by a sump and returned to the process.

Pollution Prevention and Control

a. Source Reduction- Most emissions to the atmosphere can be reduced by selection of the raw materials. Low ash sulfur requires less screening of catalyst and less solid waste. In other cases liquid sulfur filtration and SO_2 gas filtration are necessary before conversion. Yellow sulfur is preferred before other colors that have organic impurities. The latter produces a mist, which exits the absorption towers and is difficult to remove.

b.   Recycling and Byproduct Recovery- Catalyst waste is recycled or vanadium Is removed from dumped catalyst.

c.   Hazardous Materials Handling, Management, and Disposal- Sulfuric acid of 96% strength and higher may be stored in mild steel tanks. The plant areas should have impermeable flooring with all surfaces draining to a neutralization pit to deal with small acid discharges from maintenance operations.

d.   Treatment Technologies- The new sulfuric acid plants can achieve satisfactory emission levels of SO_2 and mist using a double conversion/double absorption system with high efficiency mist eliminators after the final absorption. Fiber mist eliminators having low gas velocities are used to remove all particles above 9 microns and 99% of all smaller size.

Electronic precipitators are very efficient; however, they have a high capital cost. In specific locations final tail-gas scrubbing with ammonia or caustic soda solutions is necessary.

Phosphoric Acid

Utilities

The production of 1 tonne of P_2O_5 in phosphoric acid requires 7.1 tonnes of makeup water; this includes the application process and washing requirements. The additional water may be necessary to pump the phosphogypsum slurry to the dumping site or sea. A phosphoric acid plant needs water that has two qualities: process water and cooling water. The water treatment installations do not produce harmful waste.

Waste Characteristics and Impacts

I.   Atmospheric Emissions-Phosphate rock usually contains 3%-4-5% of fluorine by weight. Hydrogen fluoride released during the acidulation of the phosphate rock combines immediately with the available silica contained in the reaction slurry. Use of low-silicate phosphate rock produces free hydrogen fluoride also.

Bothe are harmful fluoride emissions. The reactor fluorides are usually absorbed in water or in diluted solute=ions of fluosilicic acid. The modern absorbing equipment allows production of concentrated solutions, which could be processed to synthetic cryolite, aluminum fluoride, and various fluosilicic acids can be neutralized by liming. The neutralized slurry may require further treatment, such as setting, before it is discharged.

II. Wastewater- In most cases the water used to transport phosphogypsum to storage is recirculated back to the process. The phosphogypsum stacks require treatment of the run-off water for several years after the phosphoric acid plant has ceased production.

Maintaining the water in the phosphoric acid plant is an important part of the knowhow of plant operation.

Aqueous effluent from the phosphoric acid plant can originate from spills, leaking pumps, or flanges; contamination depends on the operation and maintenance standards.

III. Solid Waste- Approximately 5 tonnes of phosphogypsum dry mass is made per tonne of phosphoric acid (calculated per 1 to P2O5). When it is deposited, this waste contains about 40% water, some phosphoric acid, and trace elements from the rock: radioactive elements and heavy metals like cadmium. Other solid waste is produced during plant cleaning from scale deposits and deposits of phosphogypsum in storage tanks.

IV. Hazardous Waste- Radioactive elements and heavy metals may leach if phosphogypsum stacks are not isolated from groundwater.

V. Fugitive Emissions- Small amounts of fluorine derivatives are released into the atmosphere. Similarly there can be small amounts of phosphoric acid and sludge released during maintenance.

VI. Contaminants of Concern- All phosphate are containing traces of radioactive elements, and some contain a number of metals. These impurities are studied in relation to their impact on health and the environment. During the processing of the phosphate rock they are partitioned between ore beneficiation process waste, acid process waste, and final product.

The cadmium level in fertilizers made from phosphoric acid is increasingly causing concern, and this may influence the choice of raw materials. Igneous phosphate rock typically contains very little cadmium (<1mg Cd/ kg P), whereas sedimentary sources contain much higher but widely varying amounts (from 43 mg Cd/kg P to 380 mg Cd/kg P).

Because cadmium sulfate is water soluble and cadmium phosphate is acid soluble, most of the cadmium in phosphate rock concentrate stays in phosphoric acid and in the products made from the acid.

VII. Environmental Impact – The phosphoric acid process large amounts of phosphogypsum. The latter requires either a large land surface for disposal or a transportation system for disposal into deep sea. Disposal into the sea is controversial, and some governments established limits for this operation.

The phosphogypsum stacks should be deposited on specially lined surfaces to prevent leakage of contaminated wastewater to the groundwater. Any excess water from the stack, especially from the rain deposits, must be drained and neutralized by liming before being released.

Pollution Prevention and Control

a. Source Reduction- Most of the emissions fluorides to the atmosphere can be reduced by selection of efficient absorption equipment.

The hemihydrates (HH) process produces lower quantities of phosphogypsum and a higher concentration of phosphoric acid; however, difference in pollution problems are not substantial.

The wastewater released from the transportation of the phosphogypsum slurry must be carefully circulated.

b. Recycling and Byproduct Recovery- The proper recycling of wastewaters from the process and maintaining the water balance should be two of the technology targets.

Fluorides should be transformed into useful industrial products whenever possible. In critical cases when there is no marker for such a product, wash waters must be returned to the process and discharged with phosphogypsum.

P phosphogypsum can be processed and used as a construction material or converted to cement and sulfuric acid. The assessment of this process is given.

c. Hazardous Materials Handling, Management, and Disposal- Phosphoric acid is highly corrosive to mild steel. Much of the plant is fabricated using stainless steel, rubber-lined mild steel. Plastic piping and valving, and vessels. All storage tanks should be inside containment areas.

The plant areas should have impermeable flooring; all surfaces should drain to a neutralization pit to deal with small acid discharges from maintenance operations.

d. Treatment Technologies- Process are in commercial use for recovery of uranium from phosphoric acid.

There are presently no commercial processes for removal of cadmium from phosphoric acid; however, extensive research is carried out in view of expected stricter regulation on cadmium content.

Gaseous contaminants (fluorides) are removed by scrubbing the gases and transforming fluorides into useful products. The remaining fluorides from sewage waters and from phosphogypsum stacks are removed by two-stage lime treatment.

References

- Tuomisto, H.L.; Hodge, I.D.; Riordan, P.; Macdonald, D.W. (December 2012). "Does organic farming reduce environmental impacts? – A meta-analysis of European research". Journal of Environmental Management. 112: 309–320. doi:10.1016/j.jenvman.2012.08.018

- Schindler, David and Vallentyne, John R. (2004) Over fertilization of the World's Freshwaters and Estuaries, University of Alberta Press, p. 1, ISBN 0-88864-484-1

- Werner, Wilfried (2002) "Fertilizers, 6. Environmental Aspects" Ullmann's Encyclopedia of Industrial Chemistry, Wiley-VCH, Weinheim.doi:10.1002/14356007.n10_n05

- Bengtsson, Janne; Ahnström, Johan; Weibull, Ann-Christin (2005-04-01). "The effects of organic agriculture on biodiversity and abundance: a meta-analysis". Journal of Applied Ecology. 42 (2): 261–269. ISSN 1365-2664. doi:10.1111/j.1365-2664.2005.01005.x

- Klimaszyk, P.; Rzymski, P. (2010). "Surface Runoff as a Factor Determining Trophic State of Midforest Lake". Polish Journal of Environmental Studies. 20 (5): 1203–1210

- Kidd, Greg (1999–2000). "Pesticides and Plastic Mulch Threaten the Health of Maryland and Virginia East Shore Waters" (PDF). Pesticides and You. 19 (4): 22–23. Retrieved 23 April 2015

- Schindler, David W., Vallentyne, John R. (2008). The Algal Bowl: Overfertilization of the World's Freshwaters and Estuaries, University of Alberta Press, ISBN 0-88864-484-1

- Carpenter, S.R.; Caraco, N.F.; Smith, V.H. (1998). "Nonpoint pollution of surface waters with phosphorus and nitrogen". Ecological Applications. 8 (3): 559–568. JSTOR 2641247. doi:10.2307/2641247

- Matthews, M.W., & Barnard, S. 2015. Eutrophication and Cyanobacteria in South Africa's Standing Water Bodies: A View from Space. In South African Journal of Science. Vol. 111. No. 5/6

- Rodhe, W. (1969) "Crystallization of eutrophication concepts in North Europe". In: Eutrophication, Causes, Consequences, Correctives. National Academy of Sciences, Washington D.C., ISBN 9780309017008 , pp. 50–64

- Whiteside, M. C. (1983). "The mythical concept of eutrophication". Hydrobiologia. 103: 107–150. doi:10.1007/BF00028437

- Lawton, L.A.; G.A. Codd (1991). "Cyanobacterial (blue-green algae) toxins and their significance in UK and European waters". Journal of Soil and Water Conservation. 40: 87–97

- Angold P. G. (1997). "The Impact of a Road Upon Adjacent Heathland Vegetation: Effects on Plant Species Composition". The Journal of Applied Ecology. 34 (2): 409–417. JSTOR 2404886. doi:10.2307/2404886

- Mungall C. and McLaren, D.J. (1991) Planet under stress: the challenge of global change. Oxford University Press, New York, New York, USA, ISBN 0-19-540731-8

- Hemphill, Delbert (March 1993). "Agricultural Plastics as Solid Waste: What are the Options for Disposal?" (PDF). Hort Technology. 3 (1): 70–73. Retrieved 23 April 2015

- Duce, R A; et al. (2008). "Impacts of Atmospheric Anthropogenic Nitrogen on the Open Ocean". Science. 320: 893–89. PMID 18487184. doi:10.1126/science.1150369

- Shumway, S. E. (1990). "A Review of the Effects of Algal Blooms on Shellfish and Aquaculture". Journal of the World Aquaculture Society. 21 (2): 65–10. doi:10.1111/j.1749-7345.1990.tb00529.x

- Burkholder, JoAnn M. and Sandra E. Shumway. (2011) "Bivalve shellfish aquaculture and eutrophication", in Shellfish Aquaculture and the Environment. Ed. Sandra E. Shumway. John Wiley & Sons, ISBN 0-8138-1413-8

- Huang, Wen-Yuan; Lu, Yao-chi; Uri, Noel D. (2001). "An assessment of soil nitrogen testing considering the carry-over effect". Applied Mathematical Modelling. 25 (10): 843–860. doi:10.1016/S0307-904X(98)10001-X

- Sharpley A.N., Daniel, T.C.; Sims, J.T.; Pote, D.H. (1996). "Determining environmentally sound soil phosphorus levels". Journal of Soil and Water Conservation. 51: 160–166

- Mills, K. H.; Chalanchuk, S. M.; Allan, D. J. (2000). "Recovery of fish populations in Lake 223 from experimental acidification". Canadian Journal of Fisheries and Aquatic Sciences. 57: 192. doi:10.1139/f99-186

- Drainage Manual: A Guide to Integrating Plant, Soil, and Water Relationships for Drainage of Irrigated Lands, Interior Dept., Bureau of Reclamation, 1993, ISBN 0-16-061623-9

- Oglesby, R. T.; Edmondson, W. T. (1966). "Control of Eutrophication". Journal (Water Pollution Control Federation). 38 (9): 1452–1460. JSTOR 25035632

- Agarwal, A.; Liu, Y. (2015). "Remediation technologies for oil-contaminated sediments". Marine Pollution Bulletin. doi:10.1016/j.marpolbul.2015.09.010

- George, Rebecca; Joy, Varsha; S, Aiswarya; Jacob, Priya A. "Treatment Methods for Contaminated Soils – Translating Science into Practice" (PDF). International Journal of Education and Applied Research. Retrieved February 19, 2016

Permissions

All chapters in this book are published with permission under the Creative Commons Attribution Share Alike License or equivalent. Every chapter published in this book has been scrutinized by our experts. Their significance has been extensively debated. The topics covered herein carry significant information for a comprehensive understanding. They may even be implemented as practical applications or may be referred to as a beginning point for further studies.

We would like to thank the editorial team for lending their expertise to make the book truly unique. They have played a crucial role in the development of this book. Without their invaluable contributions this book wouldn't have been possible. They have made vital efforts to compile up to date information on the varied aspects of this subject to make this book a valuable addition to the collection of many professionals and students.

This book was conceptualized with the vision of imparting up-to-date and integrated information in this field. To ensure the same, a matchless editorial board was set up. Every individual on the board went through rigorous rounds of assessment to prove their worth. After which they invested a large part of their time researching and compiling the most relevant data for our readers.

The editorial board has been involved in producing this book since its inception. They have spent rigorous hours researching and exploring the diverse topics which have resulted in the successful publishing of this book. They have passed on their knowledge of decades through this book. To expedite this challenging task, the publisher supported the team at every step. A small team of assistant editors was also appointed to further simplify the editing procedure and attain best results for the readers.

Apart from the editorial board, the designing team has also invested a significant amount of their time in understanding the subject and creating the most relevant covers. They scrutinized every image to scout for the most suitable representation of the subject and create an appropriate cover for the book.

The publishing team has been an ardent support to the editorial, designing and production team. Their endless efforts to recruit the best for this project, has resulted in the accomplishment of this book. They are a veteran in the field of academics and their pool of knowledge is as vast as their experience in printing. Their expertise and guidance has proved useful at every step. Their uncompromising quality standards have made this book an exceptional effort. Their encouragement from time to time has been an inspiration for everyone.

The publisher and the editorial board hope that this book will prove to be a valuable piece of knowledge for students, practitioners and scholars across the globe.

Index

www.ingramcontent.com/pod-product-compliance
Lightning Source LLC
Chambersburg PA
CBHW082025190326
41458CB00010B/3277